The Story of
QUANTUM
MECHANICS

The Story of QUANTUM MECHANICS

Victor Guillemin

Drawings by Robert Guillemin

Dover Publications, Inc.
Mineola, New York

Bibliographical Note

This Dover edition, first published in 2003 and reissued in 2012, is an unabridged republication of the work originally published by Charles Scribner's Sons, New York, in 1968. Victor Guillemin, one of the author's sons, has written a new Foreword especially for the Dover edition.

Library of Congress Cataloging-in-Publication Data

Guillemin, Victor, 1896–1985.
    The story of quantum mechanics / Victor Guillemin ; drawings by Robert Guillemin.
        p. cm.
    Originally published: New York : Scribner, 1968.
    Includes bibliographical references and index.
    ISBN-13: 978-0-486-42874-1 (pbk.)
    ISBN-10: 0-486-42874-5 (pbk.)
    1. Quantum theory. 2. Science—Philosophy. I. Title.

QC174.12.G854 2003
530.12—dc21

                                                                2003043777

TO TED KEMBLE, *MENTOR*

Foreword to the Dover Edition

MY FATHER, Victor Guillemin (who died in 1985) was born in Milwaukee in 1896. He got his B.A. in physics from the University of Wisconsin in 1923 and his M.A. from Harvard in 1924. While a graduate student at Harvard, he won a Sheldon fellowship to study in Munich with Arnold Sommerfeld at the Institut für theoretische Physik; he wrote his doctoral thesis under Sommerfeld's supervision in 1926. Thus he had the good fortune to be present at the birth of quantum mechanics: Werner Heisenberg and many of the other major players in the future development of this subject, like

HUTCHINS PHOTOGRAPHY, INC.

Linus Pauling and Hans Bethe, were in Munich in the years my father was there, and, in fact, Heisenberg wrote his thesis with Sommerfeld the same year my father did. (I should mention that many of the anecdotes in *The Story of Quantum Mechanics* were witnessed firsthand during this stay in Munich.)

In the late 1920s my father wrote several well-regarded papers on quantum mechanics. However, in the 1930s he moved from physics to biophysics and spent most of his life working in experimental biophysics at Wright-Patterson Air Force base in Dayton, Ohio, and at the University of Illinois College of Medicine. On his retirement, he was asked to head up a program at Harvard whose purpose (this was just at the beginning of the post-sputnik era) was to give high-school teachers a chance to spend a year in the Harvard physics environment retooling and acquiring some familiarity with recent developments. This, I think, led him to revisit the experiences of his years in Munich and write *The Story of Quantum Mechanics*.

VICTOR GUILLEMIN
Department of Mathematics
Massachusetts Institute of Technology

Preface

PHYSICAL science, which is concerned with explaining what the universe is made of, how it is put together and how it works, has achieved two major advances in our century. One is the theory of relativity; the other is quantum mechanics. The former has somehow captured the popular fancy; and its founder, Albert Einstein, is known throughout the world. But the names of those who developed quantum mechanics are not nearly so current outside scientific circles; and what they accomplished is largely unknown to the general public. Yet quantum mechanics constitutes a deep and fundamental revision of the older classical laws of physics and presents a surprising new picture of the composition and the workings of the universe in which we live.

The province of quantum mechanics, the "new physics" as it has been called, is the sub-microscopic realm of atoms, the tiny entities of which all material substance is composed. Although this atomic substratum is not immediately apparent to our senses, its activities exert their effects upon many phenomena that can be observed. They influence the course of events in the remote stars of the heavens, in our earthly environment, in all living creatures, including ourselves.

When, in the early years of our century, physicists first developed new experimental techniques for investigating the structure and activity of atoms, it soon became apparent that the laws of nature which had been developed while dealing with familiar things that could be seen and handled were not adequate to explain the new findings. Novel concepts and theories had to be developed; and these were so surprising, so contrary to seemingly self-evident tenets, that they came to be accepted only reluctantly under the force of convincing experimental evidence.

To tell of the ways in which the ingenious new experiments in atomic physics were devised, what they revealed, and how this new knowledge was finally accounted for after much hard work, many tentative hypotheses and a few brilliant flashes of insight—this is the story of quantum mechanics.

The subject matter of this book falls naturally into four divisions. Chapters 1 through 5 are introductory. They trace the development of physics, primarily of ideas regarding the nature of matter and of light, from antiquity through the period of classical science in the seventeenth, eighteenth and nineteenth centuries, and on into the first two decades of the twentieth. Chapters 6 through 8 explain how quantum mechanics then developed from what had gone before, how it coped with previously unsolved problems and how it gave an excellent account of all observable atomic phenomena. Chapters 9 through 14 are devoted to matters which are presently at the forefront of research, matters pertaining to the "elementary particles," presumably the ultimate constituents of atoms and thus of all material things. Finally, in Chapters 15 through 18, the concepts of physical science, both the old and the new, are brought to bear on questions, some of ancient origin, in the realm of philosophy and religion. In these discussions, the controversial ideas of causality, determinism and free will, are given careful consideration.

I had the good fortune of attending the *Institut für theoretische Physik* at the University of Munich at the time when quantum mechanics was in its early stages of development. There the seminar of Professor Arnold Sommerfeld, a leading figure in theoretical physics, was attracting a group of scientists doing research in this new field. My association with this work engendered an enduring interest which has led, after many years, to the writing of this book. It is the consequence of innumerable discussions, both formal and informal, of hundreds of lectures attended and other hundreds delivered, of reading books and scientific journals, and, in no small measure, of many years given to the teaching of physics during which numerous ideas were sharpened and clarified by the searching questions of my students.

It is thus impossible for me to acknowledge all the sources from which I have drawn. Some of the more recent books and articles which I have consulted are referred to specifically at appropriate points in the text and others are listed in the annotated bibliography.

High praise is due my wife who typed and retyped the various rewritings of the manuscript with great patience and contributed, by many small changes, to its smooth readability. Any gaucheries that remain are, of course, my own.

This book is intended to be read by those who, though lacking formal training in science or mathematics, maintain a lively interest in the ways by which scientists are probing the workings of nature. It is written to convey sound factual information, some of which may well be found surprising. Beyond the presentation of scientific facts, concepts and theories I have tried to convey a feeling for the ways in which professional scientists work and think.

Finally, I confess that I wrote the book because I enjoyed doing it. I can but hope that my readers will obtain their share of pleasure in becoming aware of its contents.

Contents

Figures and Tables

TABLES

Portraits

The Story of
QUANTUM
MECHANICS

1

The Science of Motion

Ours is a universe of incessant motion. Motion is everywhere, in the celestial realm of the stars, in our familiar earthly environment, in the tiny invisible atoms of which all material things are made. So long as life persists in our bodies, our hearts and lungs and viscera never cease their activity. An idle thought flitting through our consciousness means that somewhere a chemical substance has crossed the threshold of a nerve cell. Whenever anything at all happens in the universe, something moves.

The science that deals with motion is called *mechanics*. It pertains to all motions of all things, to the forces that alter motions and the energy changes associated with them. These basic concepts of mechanics, *motion, force* and *energy,* are pertinent to all the sciences, both in the realms of living organisms and of inanimate objects.

It is thus understandable that throughout all of history, from ancient times to the present, general progress in science has been linked in some degree to progress in mechanics. This is true most notably of the events, beginning around the year 1600, which ushered in that grand forward sweep of knowledge, progressing through the seventeenth, eighteenth and nineteenth centuries, which historians have called the *classical period of science.* These developments brought about a veritable revolution in natural philosophy, leading from the mysticism and obscurities of medieval *scholasticism* to the rational spirit of modern science.

NEWTONIAN MECHANICS

THE scientific revolution of the seventeenth century was dominated by the genius of one man, Sir Isaac Newton (1642-1727), rightly acclaimed as one of the greatest intellects of all times. He received notable honors during his lifetime, and he was buried in Westminster Abbey with ceremonies which moved the French writer Voltaire to remark that the English accord eulogies to scientists which other nations reserve for kings.

Newton made epochal discoveries in various branches of science and mathematics, but of particular pertinence to our story is his work in mechanics. Throughout the great creative era of classical science Newtonian mechanics helped to shape the course of events. Thus, its essence needs to be understood as the point of departure for the new systems of mechanics that arose in the twentieth century.

Natural philosophers of the Middle Ages had based their ideas about motion on the works of ancient Greek savants, notably on those of Aristotle (384-322 B.C.), pupil of Plato and teacher of Alexander the Great. Aristotle attributed all natural motions to innate tendencies of the moving bodies—an almost animistic conception. A stone tossed into the air, for example, hurries back to earth because of a predisposition to return to its "natural place."

In a very basic matter concerning motion, Aristotle was quite wrong. He maintained that the natural state of all bodies is the *state of rest*; if a body is made to move by a push or pull, it quickly reverts to its natural state of rest when left to itself. While this idea seems in accord with common experience—witness a cart that has been given a push—Newton saw clearly that the cart is not left to itself; rather, it is left to the action of friction forces, which quickly rob it of its motion. Seeing that, where friction is minimized as in sliding over smooth ice, motion persists for a longer time, he divined that a body freed of all forces should continue to move forever at constant velocity along a straight line. (This is the import of Newton's *first law* of motion.) Forces, he saw, are needed, not to maintain constant motion, but to *change* it, to start or stop motion, to speed it up or slow it down or to change its direction. (The relation of force to change of motion is the burden of the *second law*.)

In addition to being wrong in this and other important matters, Aristotelian mechanics was quite useless for predicting the motions that should occur under given conditions or for learning how to control them. By contrast, Newton's clear and simple laws lent themselves ideally to definitive mathematical analysis of motions and of their relations to forces. Newton built so well that his laws continued to serve, for nearly two hundred years after his death, as the basis for all further developments in mechanics, both in disciplines that pertain to inanimate objects and in those that deal with the dynamics of living things.

Of equal import was Newton's discovery of the *law of universal gravitation*. The anecdote about the falling apple has become a legend; but unlike many others, this is well authenticated. While staying at his mother's farm to escape the plague that had appeared in London, he observed an apple falling from a tree and got to musing about the earth's downward force of gravitation, which caused it to fall. Knowing that this force extends to the highest mountain tops, he was moved to speculate whether it might not reach even to the enormous height of the moon and account for the motion of that satellite in its orbit around the earth. Making a definite mathematical assumption about the nature of the gravitational force, namely, that it diminishes in inverse proportion to the square of the distance from the earth's center, he was able to compare the force on the moon and on the apple and, using his second law of motion, to calculate how the moon should move under the influence of this force. He was gratified to be able to jot down in his private notes that the astronomical observations and the results of his calculations "answer pretty nearly."

Extending his calculations to the motions of the planets in their far-flung orbits about the sun, he again obtained excellent agreement with astronomical measurements. Thus, at one stroke, he swept aside all the erroneous notions, held over a period of twenty centuries, about an immanent difference between things on earth and in the heavens. Where all others had failed, he succeeded in constructing a rational celestial mechanics; and he showed that the motions of both celestial and terrestrial bodies are governed by the same laws.

Newton indicated this clearly in a diagram much like the one shown in Figure 1. If a projectile is merely dropped from the top

Figure 1. Newton's diagram showing the similarity of terrestrial and celestial motion.

of a cliff, it lands at the base at point *A*. But if it is propelled forward, it falls along a curved path to point *B*. As the initial forward velocity is increased, the projectile passes farther around the curve of the earth and strikes at successively more remote points *C*, *D*, or *E*. Finally, at a sufficiently high initial velocity it does not fall at all but continues to circle on path *F*, round and round the earth, much in the manner of the moon. Newton knew, of course, that this idea was hypothetical; actually, the friction of the air would soon reduce the projectile's speed and bring it back to earth. He certainly would have been astounded to learn that less than three hundred years thence projectiles (artificial, rocket-propelled satellites) would actually be circling the earth high above the atmosphere.

Of himself Newton said: "If I have seen farther than other men, it is because I have stood on the shoulders of giants." This is true in an important sense. Newton could not have done his work nor could he have received recognition for it, if other men had not created the necessary preconditions through advancing knowledge and, more importantly, by establishing a favorable climate.

There was Francis Bacon (1561-1626), courtier to Queen Elizabeth and to her successor, James I. Bacon was a philosopher, statesman and diplomat, to whom at one time the authorship of Shakespeare's plays was erroneously attributed. He elaborated a thorough experimental philosophy for science, which was a radical departure from the prevalent conception that knowledge may be augmented through rhetorical argumentation based on the teachings of ancient and sanctioned authorities. Through his influence he made scientific experimentation popular among prominent men, a circumstance which led eventually to the founding, on Newton's twentieth birthday, of the renowned *Royal Society of London for Improving Natural Knowledge by Experiment*, usually known simply as the "Royal Society."

Then there was René Descartes (1596-1650), French philosopher, theologian, mathematician and scientist, who asserted that the entire physical universe is a great machine operating according to laws that may be discovered by the power of human reason, particularly when it is disciplined by the clear logic of mathematics. His own prowess in that field led him to exaggerate the importance of mathematical reasoning and to deprecate experimentation.

The importance of careful experimentation observation was exemplified, however, in the lives of two other "giants." One, the German Johannes Kepler (1571-1630), spent many years poring over the precise astronomical records of planetary motion set down by his teacher, the Danish astronomer Tycho Brahe (1546-1601), and summarizing these observations in a manner ideally suited to serve later as a verification of Newton's celestial mechanics. And there was the Italian Galileo Galilei (1564-1642), who made important observations on the acceleration of rolling metal spheres under the influence of the earth's gravitational force and carefully drew valid conclusions about motions of terrestrial bodies.

Newton was indeed indebted to these men, but he surpassed

them all. He embraced the experimental philosophy of Bacon and added to it a transcendent scientific intuition. His laws of mechanics are a perfect paradigm of Descartes's mathematical laws of nature. But in accord with his famous dictum, "Hypotheses non fingo" (I make no hypotheses), Newton based these laws, not on unsupported mathematical speculations, but on sound experimental evidence, in the manner of Kepler and Galileo.

THE EXPERIMENTAL-DETERMINISTIC PHILOSOPHY

OF great moment for the course of intellectual progress was the philosophy engendered by the advance of classical science. Descartes had asserted that the universe is a great machine, and Newton had formulated the laws that govern its operation. In a machine everything that occurs is the direct consequence of what has happened before. Each event is caused by those that have come before and is, in turn, the cause of subsequent events. Thus, the laws of nature appear to be *deterministic*; they order the sequence of natural phenomena into an unbroken chain of *cause* and *effect*.

Moreover, a machine is essentially observable and comprehensible. It contains no elements, however complex, that transcend scrutiny. Consequently, within the scope of classical scientific philosophy, the *data* obtained by observation and experiment should be the source of all that can be known about the physical universe; and speculations about entities that lie beyond the scope of experimental verification are vain and unnecessary.

Actually, matters did not not proceed quite in accord with this philosophy, for a human enterprise of vast scope cannot be held to a pat formula. In the sciences that pertain to living organisms, determinism was challenged repeatedly and with great heat, particularly in its relation to human affairs. For the physical sciences, however, it remained the indispensable prerequisite for all scientific work; and it held undisputed sway well into the twentieth century.

As was to be expected, scientists of the classical period did not always refrain from imaginative conjectures extending into realms quite inaccessible to experiment, in spite of the protestations of eminent men against unseemly speculation. As things turned out, some of these hypotheses spurred on experimenters to devise new and rewarding procedures that opened up fresh areas of knowledge.

In the second half of the nineteenth century certain important developments in electricity and in optics engendered grave doubts concerning the possibility of describing the physical universe wholly as a mechanical contrivance. But in the realm of mechanics the conception of a machinelike universe was adhered to almost without exception throughout the entire course of classical science.

THE TWENTIETH CENTURY

Our present century has witnessed two major extensions of mechanics into areas which were not accessible through the experimental methods available in Newton's day. One of these is the *theory of relativity*. The other is *quantum mechanics*.

The theory of relativity has wrought profound changes in our concepts of space and time and of the relation between mass and energy; and it has enhanced in great measure our understanding of the physical universe. With regard to the laws of motion, the new relativistic mechanics is important in dealing with objects which move at velocities comparable with that of light in free space (186,000 miles per second). Such extreme velocities have been observed in recent times by astronomers studying remote galaxies; and they have been actually achieved for very tiny particles such as electrons in the laboratories of the physicists. For motions in the world of everyday experience, and even for the impressive velocities of space vehicles measured in thousands of miles per hour, the laws of Newtonian mechanics are still quite adequate.

Quantum mechanics is concerned with the submicroscopic realm of the atoms, the tiny entities of which all material things are constituted. In the early years of our century when physicists first succeeded in devising ingenious new experiments to probe the structure and the activity of the atoms, it soon became apparent that the laws of classical physics, which had been developed by working with large, directly observable objects, could not account for the new findings. Even concepts which had been thought to be true as a matter of course were found to be no longer tenable.

That such pronounced changes in the nature of things could result from a mere change of scale was certainly not to be expected. Whether the atoms were pictured as tiny grains smaller than specks of dust or as more complex entities with internal moving parts, it had been assumed that the natural laws pertaining to their activities should be the same as those which determine the course of events among large, familiar objects. For engineers conduct experiments with scaled-down models of ships and aircraft to obtain reliable information about the performance of the full-size ones. And the readers of Jonathan Swift's engaging tales about Gulliver's travels have no difficulty in supposing that the tiny inhabitants of the land of Lilliput could actually exist; all of which is in accord with the idea that "large" and "small" are relative terms. There is, in fact, the classical scientific "principle of similitude" which asserts that the course of events is, quite generally, independent of absolute scale, as expressed rather whimsically in the doggerel:

> Larger fleas have smaller fleas
> Upon their backs to bite 'em
> And these in turn have lesser fleas
> And so *ad infinitum*!

Yet physicists found most convincing experimental evidence that a millionfold downward plunge in size, from that of familiar objects, results in profound alterations in the interplay of forces, motions and energies, quite incompatible with classical laws. Their observations implied for example that the state of an atom, its structure and its energy, can vary by abrupt, discontinuous transitions which cannot be described as happening in space and time nor can they be pictured as a cause-and-effect sequence. The fact that such novel behavior might result from a mere change of scale could however be given logical justification by relating it to an utterly new entity appearing in quantum mechanics, a "matter wave" associated with moving objects, which could serve as a gauge by which size, or rather mass, acquires definitive significance.

But further, as the new developments came to be better understood, the remarkable comprehension emerged that there is actually no contradiction between the laws of quantum mechanics and those of Newton. Rather, it appeared that the old laws are included within the more comprehensive new ones, in the sense that the former have only a limited scope while the latter appear to be valid over the whole range of size, from atoms on up to objects in our familiar environment and even beyond into the vast astronomical space.

Expressed differently, when quantum mechanics is used to deal with events among the atoms, it yields new information which the cruder classical mechanics could not provide. But when the new mechanics is applied to the gross motions of large objects it gives results which are, to all appearances, the same as those of the old. This could hardly be otherwise. For quantum mechanics would certainly be untenable if it did not agree with the laws of classical mechanics in the areas for which these laws have been thoroughly verified by experiment.

It does not follow, however, that quantum mechanics is of no real significance in the world of everyday affairs. For the activities in the atomic substratum manifest themselves in a great variety of phenomena which are directly observable. Thus the light radiated by the stars is bound up with atomic interactions going on in their intensely hot interiors. And the special characteristics of each kind of living organism is determined by a genetic code written in characters of atomic size.

It is hardly surprising that a scientific advance of such broad scope should have found important applications in technology. The special electronic devices, for example, which flowed from the new knowledge about atoms and electrons, were essential in bringing computers and space vehicles to their present stage of development. But perhaps of greater moment is the fact that quantum mechanics provides an incisive challenge to the mechanistic, deterministic philosophy of classical science. This is one of the reasons why it has had repercussions in disciplines far removed from science, in metaphysics, ethics and theology. These are matters which will be examined in considerable detail after their scientific background has been discussed.

Obviously the story of quantum mechanics includes the atoms. It is appropriate therefore to consider at some length how scientists arrived at their present concept of atomic structure and dynamics. And since much of this has been learned by observing the interaction of atoms and light rays, it will be necessary to digress at appropriate points to discuss the nature of light.

2

The History of Atomism

As everyone has heard *ad nauseam,* we are living in the "atomic age." To the general public this is evidenced by such spectacular events as the explosion of atom bombs or the propulsion of ships by atomic power. For physical scientists the beginning of the atomic age goes farther back, to the turn of the century, when first they gave serious attention to the detailed study of atoms.

Atoms are not, however, an invention of the twentieth century. The idea that the material substance of the universe is constituted of tiny, distinct particles was proposed by the Greek philosopher Leucippus about 450 B.C. and was developed several decades later in considerable detail by Democritus of Abderra (470-380 B.C.). In the latter's philosophy the very fact of being means to exist as atoms; *the atoms* and the *void* in which they move ceaselessly comprehends all of existence.

Democritus describes the atoms as being so minute, so far below the limit of visibility, that even a tiny speck of matter contains a huge number of them. They are all made of the same basic substance, but there are many kinds, differing in size and shape. Most things are assumed to be mixtures of several kinds of atoms, the numerous varieties of materials observed in the universe being just the various possible combinations and arrangements of these atoms. All perceived changes in things are brought about by alterations of these groupings. The atoms themselves (the name is derived from the Greek word *atomos,* meaning "indivisible") are ultimate entities, *unchangeable, indestructible, eternal.*

For these conceptions there was, of course, no factual evidence. Democritus was concerned primarily with the philosophical problem relating to the existence of both permanence and change in the nature of things. The assumption of one basic substance common to all atoms achieved simplicity and uniformity; and in a world where all things are subject to change and death and decay, the concept of indestructible atoms provided a comforting substratum of permanence.

Throughout the Middle Ages the atomism of Democritus was eclipsed by a rival concept which had the support of Aristotle's great authority. This concept, that all material things are composed of four "elements," *earth, water, air* and *fire,* was first proposed by the Greek statesman, poet and philosopher Empedocles (490-430 B.C.) and was adopted and elaborated by Aristotle. The latter related each of the four elements to the sense of *touch,* which he thought to be the most basic of sense perceptions. In particular, he associated the impressions of wet and dry and of hot and cold with the four elements, according to the following scheme: earth—cold, dry; water—cold, wet; air—hot, wet; fire—hot, dry.

Over a span of some twenty centuries the doctrine of these elements dominated all philosophical discussions bearing on the nature of matter. But the atoms of Democritus were never entirely forgotten; and as the dawn of rational classical science approached, they were again much in the thoughts of natural philosophers. Francis Bacon discussed atoms in his scientific writings; and while he disapproved, in general, of any ideas not derived directly from experiment, he did concede that "atomism is applicable with excellent effect to the exposition of nature."

THE CHEMICAL ATOMS

The concept of discrete bits of matter moving incessantly in a void lent itself well to the mechanistic philosophy of classical science. Newton himself supported this idea wholeheartedly but was careful to propose it as a reasonable conjecture, not as a conclusion based upon experimental evidence. In his monumental *Optics,* published in 1704, he wrote:

> All things being considered, it seems probable to me, that God in the beginning formed matter in solid, massy, hard, impenetrable, moveable particles, of such sizes and figures, and with such other properties, and in such proportion to space, as most conduced to the end for which he formed them; and that these primitive particles, being solids, are incomparably harder than any porous bodies compounded of them; even so very hard as never to wear or break in pieces; no ordinary power being able to divide what God himself made one in the first creation. While the particles continue entire, they may compose bodies of one and the same nature and texture in all ages. But should they wear away, or break in pieces, the nature of things depending on them would be changed. Water and earth, composed of old worn particles and fragments of particles would not be of the same nature and texture now with water and earth composed of entire particles at the beginning. And therefore, that nature may be lasting, the changes of corporeal things are to be placed only in the various separations and new associations and motions of these permanent particles.

The concept of minute, indestructible atoms persisted throughout the two following centuries, gaining support from an increasing body of experimental evidence. Even during Newton's lifetime the English scientist Robert Boyle (1627-91) made substantial progress toward an atomic concept of material substance. His important work, *The Sceptical Chemist,* published in 1661, marks the transition from alchemy to the science of chemistry. In this book he gave an experimental definition of a *chemical element* as a substance that cannot be separated into simpler ones by chemical means. Substances that can be thus separated, and that can also be produced by combining their constituent elements chemically, he called *chemical compounds.* In the language of atomism an element is a substance made up of atoms of one sort only while compounds contain two or more kinds in combination.

This atomic composition was described more explicitly and

was given its first sound experimental support in the work of the English chemist John Dalton (1766-1844). He used the precise evidence of the chemical balance to support his ideas about the atomic composition of the matter. In the two compound substances carbon monoxide and carbon dioxide, for example, the proportions by weight of oxygen and of carbon are, respectively, four to three and eight to three. This may be explained by assuming that the weights of single atoms of oxygen and of carbon stand in the ratio of four to three, and that a "compound atom" of carbon monoxide is a closely bound group of one oxygen and one carbon atom while carbon dioxide is composed of groups of three atoms, two of oxygen and one of carbon. In modern chemical symbols the structure of these two groups, called *molecules*, is indicated as CO and CO_2. By careful measurements of the combining weights in many compounds, Dalton was able to determine values for the relative atomic weights of many of the elements known in his day.

Like Newton, Dalton was a firm believer in atoms, which he pictured, however, as light, fluffy balls equipped with tiny hooks to hold them together in compounds. One hundred years after Newton he wrote:

> Matter, though divisible to an *extreme degree,* is nevertheless not infinitely divisible. . . . The existence of ultimate particles of matter can scarcely be doubted though they are probably much too small ever to be exhibited by microscopic improvements.

Succeding generations of scientists continued the study of matter by chemical techniques, gradually adding to the list of known elements and determining their atomic weights with increasing precision. Innumerable new compounds were studied, including some of the very complex ones, the *organic compounds*, that make up the tissues of living organisms. Thus, the ways in which all the various kinds of atoms interact to form the molecules of chemical compounds came to be known in great detail.

All of this information was set in good order by two men, the Russian Dimitri Ivanovich Mendeleev (1834-1907) and the German Julius Lothar Meyer (1830-95). They published independently and almost simultaneously a general scheme of organization, known as the *periodic table of elements,* in which all the known elements are arranged according to their atomic weights and chemical properties. Mendeleev left places in his table for certain as yet unknown elements and boldly predicted what they would be like when and if they should be discovered. When three of these were actually identified shortly thereafter and found to have just the predicted characteristics, Mendeleev achieved worldwide recognition; and his table of classification of the elements was universally accepted. A modern periodic table of elements is shown on page 18.

THE MOLECULES OF THE PHYSICISTS

WHILE these developments in chemistry were being made, physicists were contributing additional evidence for the reality of tiny, discrete particles of matter. This came about largely through the study of gases, the most tenuous form of matter, in which the individual particles are presumably well separated and thus most apt to reveal their individual properties. While chemists studied the atoms and the ways in which they combine to form molecules, physicists were interested in the molecules themselves, since they are physically the individual particles of gases, the particles of carbon dioxide gas being, for example, the molecules CO_2.

It is a common property of gases that they exert a *pressure* upon the walls of any container in which they are confined, as in an automobile tire. Newton attributed this pressure to a force of repulsion acting between the molecules, which causes them to press upon the walls of the confining vessel. This idea was found to be untenable; and it remained for the Swiss mathematician Daniel Bernoulli (1700-82) to suggest the true explanation of gas pressure. He assumed that the molecules of a gas are in perpetual rapid motion and that they collide with one another and with the walls of the container many billions of times in a second, so that these myriad impacts act, in effect, as a steady pressure on the walls. This idea was elaborated by two scientists, the Scottish physicist James Clerk Maxwell (1831-79) and the Austrian Ludwig Boltzmann (1844-1906), into the *kinetic theory of gases*. Since they had to deal with the chaotic flight of the molecules,

Table 1. Modern periodic table of elements.*

I	II	III	IV	V	VI	VII	VIII			0
1 H 1.0080										2 He 4.003
3 Li 6.940	4 Be 9.013	5 B 10.82	6 C 12.011	7 N 14.008	8 O 16.000	9 F 19.00				10 Ne 20.183
11 Na 22.991	12 Mg 24.32	13 Al 26.98	14 Si 28.09	15 P 30.975	16 S 32.066	17 Cl 35.457				18 Ar 39.944
19 K 39.100	20 Ca 40.08	21 Sc 44.96	22 Ti 47.90	23 V 50.95	24 Cr 52.01	25 Mn 54.94	26 Fe 55.85	27 Co 58.94	28 Ni 58.71	36 Kr 83.80
29 Cu 63.54	30 Zn 65.38	31 Ga 69.72	32 Ge 72.60	33 As 74.91	34 Se 78.96	35 Br 79.916				
37 Rb 85.48	38 Sr 87.63	39 Y 88.92	40 Zr 91.22	41 Nb 92.91	42 Mo 95.95	43 Tc 99	44 Ru 101.10	45 Rh 102.91	46 Pd 106.4	54 Xe 131.30
47 Ag 107.880	48 Cd 112.41	49 In 114.82	50 Sn 118.70	51 Sb 121.76	52 Te 127.61	53 I 126.91				
55 Cs 132.91	56 Ba 137.36	57–71 La series**	72 Hf 178.50	73 Ta 180.95	74 W 183.86	75 Re 186.22	76 Os 190.2	77 Ir 192.2	78 Pt 195.09	86 Rn (222)
79 Au 197.0	80 Hg 200.61	81 Tl 204.39	82 Pb 207.21	83 Bi 209.00	84 Po 210	85 At 210				
87 Fr (223)	88 Ra 226.05	89– Ac series***								

** Lanthanide series:

57 La 138.92	58 Ce 140.13	59 Pr 140.92	60 Nd 144.27	61 Pm 145	62 Sm 150.35	63 Eu 152.0	64 Gd 157.26	65 Tb 158.93	66 Dy 162.51	67 Ho 164.94	68 Er 167.27	69 Tm 168.94	70 Yb 173.04	71 Lu 174.99

*** Actinide series:

89 Ac 227	90 Th 232.05	91 Pa 231	92 U 238.07	93 Np 237	94 Pu 242	95 Am 243	96 Cm 246	97 Bk 248	98 Cf 251	99 Es 253	100 Fm 255	101 Md 256	102 No 256	103 Lw 257

*The number to the left of each symbol is the *atomic number*, which is equal to the number of electrons in the atom; the number below each symbol is the *chemical atomic weight*, which is the average of the atomic weights of the various isotopes of that element.

they made use of the mathematical *theory of probability,* which governs random events. This branch of mathematics was invented by the French mathematician, physicist and philosopher Blaise Pascal (1623-62) at the instigation of a friend who wished to know whether there might be a mathematical explanation for his constant losses at the dice table.

The kinetic theory, which pertains to *thermodynamics* (the branch of physics that deals with the relations of heat and mechanical energy), yielded the important new insights that, as substances get hotter, their molecules move more rapidly, and that heat is a form of energy: the energy of this molecular motion. But, of even greater interest, this theory affords a striking example of the fact that chaos, provided it is complete chaos, can produce, on the average over large numbers of individual events, uniform and predictable results. Thus, the very fact that the flight of the molecules is completely at random assures that the gas pressure, the average result of many random impacts, is uniform throughout a gas-filled vessel.

As the nineteenth century drew to a close, chemists had discovered all but a few of the elements in Mendeleev's table, which, with the complete list of the ninety-two naturally occurring elements, ranges from hydrogen, with the lightest atoms, to uranium, with the heaviest. Physicists had contributed additional information about the motion and energy of these smallest parts. Thus, it seemed the age-old quest after the nature of matter had been brought to a close.

THE COMING OF THE ATOMIC AGE

ABOUT two decades before the turn of the century Edwin Hall (1855-1938), who was later to become a renowned physicist, presented himself to Professor John Trowbridge (1843-1923), head of the physics department at Harvard University, and asked to be accepted as a graduate student. He was told, not unkindly, that he would be well advised to choose another career because physics was a field of inquiry in which nothing much of importance was left to be done! The question of how a highly knowledgeable person could have been so utterly mistaken in an area in which he had outstanding competence certainly demands an answer, particularly since this opinion of Professor Trowbridge was shared by many of his contemporaries. The explanation lies in a conception of the proper methods and aims of the physical sciences that was commonly held in the closing years of the nineteenth century.

Expressed in broadest terms, the task of the physical sciences is to determine what the universe is made of and to describe how it works. In accord with the experimental philosophy of classical science, the job of investigating the *stuff* of the universe is accomplished by isolating each individual pure substance and observing and measuring all its specific properties, mechanical, chemical, thermal, optical, electrical and all the rest. This is, of course, an endless job. One can always find new substances and can conceive of additional properties to measure, and any existing measurement may be carried to a higher degree of precision. But this kind of work had been pursued with great diligence and skill throughout the eighteenth and nineteenth centuries. Thus, Professor Trowbridge felt justified in his opinion that nothing new or exciting could come of it.

Concerning the task of describing *how the universe works*, again the classical approach is by way of observation and experiment. When many observations are made in a given area, it frequently happens that they exhibit a distinct pattern; certain regularities emerge which point the way to more extended observations. When such a comprehensive body of data is contemplated imaginatively, it can perhaps be summed up in one concise statement, frequently in the form of a mathematical relation, which is called a *law of nature*. A natural law is thus an assertion about what nature has been *observed* to do, not what it is *compelled* to do. The laws of Newtonian mechanics are an excellent example of broadly comprehensive natural laws.

Throughout the two centuries following Newton's lifetime, a succession of eminent scientists arranged a vast corpus of experimental observations in mechanics, optics, thermodynamics and electromagnetism into mathematical laws of great breadth and power. It was known that there were still a few "clouds" on the otherwise clear horizon of science, groups of observations that stubbornly refused to comply with the classical laws. But it was generally agreed that the same scientific procedures which had already wrought such an impressive record of success could cope with these apparently minor difficulties in due time.

This was the widely held opinion which led Professor Trowbridge to conclude that the physical sciences had well-nigh completed their task of describing what the universe is made of and what happens in it.

Taken as a whole, classical physics is concerned primarily with the *description* of things rather than with *explanations*. The classical laws describe *how* things behave but contain no hint of *why* they do so. Similarly, the characteristics of material substances are described in great detail but no explanation is advanced for them—why, for example, glass is transparent and copper is opaque, why copper permits the flow of electric current while glass does not. Newton's famous dictum, "Hypotheses non fingo," was the guiding principle; and explanation, the formulating of hypotheses, was not considered an essential function of science.

The distinction between description and explanation in science is by no means unequivocal. An explanation may be merely a detailed description at a more fundamental level. And since all material substance is constituted of atoms, any attempt at a deeply penetrating explanation for the nature of things must necessarily come to deal with the characteristics of these atoms.

But here a conflict arose with the experimental philosophy of classical physics. Speculations about the atoms had always implied that they are far below the limit of direct observation and measurement, even by the most powerful microscope and the most sensitive chemical balance. All actual measurements reveal only the phenomena that might be attributed to the gross average behavior of huge numbers of these minute entities. Observed chemical reactions proceed *as if* they are due to the regrouping of individual atoms; the pressure of a gas acts *as if* it could be attributed to the swarming of tiny molecules.

To explain observable phenomena in terms of ideas that are beyond experimental verification appeared to many thoughtful scientists a highly dubious procedure. The German chemist Friedrich Wilhelm Ostwald (1853-1932) cautioned his fellow scientists against "hypothetical conjectures that lead to no verifiable conclusions." And the American physicist Josiah Willard Gibbs (1839-

1903) wrote: "He builds on an insecure foundation who bases his conclusions on ideas concerning the ultimate structure of matter," meaning ideas concerning atoms.

Fortunately, there were more venturesome souls. Boltzmann, among others, was convinced that the concept of atoms had become indispensable in the physical sciences. There were even occasional speculations about the *structure* of atoms. As early as 1815 the English physician William Prout (1785-1850), who first thought of dividing foods into proteins, fats and carbohydrates, conceived the idea that the atoms of the various elements might be closely bound clusters of hydrogen atoms, the lightest of all atoms. He himself considered this idea so hypothetical that he published it anonymously. For there was no experimental evidence whatsoever to refute the conception of atoms as the ultimate entities of matter, the hard, indestructible particles of Democritus. Prout's hypothesis, as well as later speculations about atomic structure, remained mere speculation.

Then, just before the dawn of the twentieth century, there occurred a number of unforeseen events which ushered in a whole new era of scientific research, events which will be discussed later in some detail. Completely new techniques of observation were discovered, and these brought forth the exciting promise that the properties of individual atoms might after all be amenable to experimental investigation.

Atoms suddenly seemed very real, and physicists dared to look forward hopefully toward understanding the physical universe in terms of its constituent atoms. This hope has proven to be well founded. Since the beginning of our century the program of relating experimental observations to the properties of atoms has been proceeding with ever-increasing success. In this all-pervasive sense science entered upon the atomic age.

As they began to deal with the problems of atoms, physicists used the classical laws, which had been developed over the centuries through experience in working with objects in the familiar large-scale world, for there were no other laws. As said before, it soon became apparent, however, that these laws were not adequate to cope with phenomena at the minute scale of the atomic world. Deductions based on classical laws simply did not agree with experimental observations. Time and again *ad hoc* modifications had to be introduced, some of which were quite inconsistent with the basic principles of classical science. Thus, atomic theories became a distressing patchwork of contradictory ideas.

Then, in the third decade of our century, a novel and trenchant concept was formulated which held the solution to these difficulties. This concept formed the basis for a completely new kind of mechanics, *quantum mechanics,* which proved extraordinarily successful in dealing with atomic phenomena. The contradictions with classical laws were resolved when it became evident that these older laws are but approximations to the more comprehensive quantum-mechanical laws, approximations adequate to deal with gross, large-scale phenomena but not with the minute details encountered in the realm of atoms. But before discussing these new developments, it is necessary to consider the background of knowledge about atoms from which they arose.

3

Atomic Structure

SIZE, MASS AND MOTION

CONJECTURES about the *size* of atoms, which the ancients had described merely as extremely minute, go back to the very beginning of classical physics, when measurements and numerical values were given much emphasis. Newton himself made estimates of atomic size by considering such things as soap bubbles and very thin gold foils, which, he thought, should consist of at least several layers of atoms. Others made similar guesses based on estimated sizes of smoke particles and other minute things. These speculations led to atomic sizes of the order of a hundred-thousandth of an inch. Through later improved measurements of various kinds these estimates were constantly revised downward.

A good value of molecular sizes was obtained about 1860 by Maxwell in his theoretical studies of gases. He derived mathematical formulas relating some of their measurable properties, such as the rate at which one gas diffuses through another, to the size of the molecules. He was able to calculate for the hydrogen molecule a size of about 0.000,000,05 centimeter (there are about two and one-half centimeters to the inch).

Just as it is a simple matter to compute the number of cherries in a basket of known size when the size of one cherry is known, so it is possible, given the size of a single molecule, to find the number of them in a bit of substance big enough to be measured. Since the bit may also be weighed, one can determine the weight of a single molecule, just as the weight of one cherry may be found by weighing a basket containing a known number of them. In this manner the weight of a single hydrogen molecule was found to be close to 0.000,000,000,000,000,000,000,003,3 gram (a gram is one one-twenty-eighth ounce), from which it follows that there are about 300,000,000,000,000,000,000,000 of them in a gram of hydrogen. Measurements of various other properties of gross matter, which could be related to molecular size and weight, gave similar values.

Scientists are frequently confronted with very large and very small numbers, such as these, that when written with strings of zeroes are awkward to manipulate. They have therefore developed the *scientific notation*, using powers of ten, the *power* being a small number attached to the ten to show how many times it is to be multiplied by itself. For example:

$$10^1 = 10$$
$$10^2 = 10 \times 10 = 100$$
$$10^3 = 10 \times 10 \times 10 = 1000$$

Any large number may be written in terms of powers of ten. For example:

$$80,000 = 8 \times 10,000 = 8 \times 10^4$$
$$1,600,000 = 16 \times 10^5 = 1.6 \times 10^6$$

The latter form (a number between zero and ten multiplied by a power of ten) is the usual notation. In this manner the number of hydrogen molecules per gram (three hundred thousand million million million) is written simply 3×10^{23}.

Small numbers are written in *negative* powers of ten which indicate fractions. For example:

$$10^{-1} = 1/10 = 0.1$$
$$10^{-2} = 1/100 = 0.01$$
$$10^{-3} = 1/1000 = 0.001$$

In this notation, one-millionth is 10^{-6}, one-millionth-millionth is 10^{-12}, and the weight of the hydrogen molecule (three and three-tenths millionth-millionth-millionth-millionth) is 3.3×10^{-24} gram. Maxwell's value for molecular diameters is 5×10^{-8} centimeter.

Since hydrogen gas consists of diatomic molecules (H_2), the weight of one atom is just half that of the molecule. Once the weight of the hydrogen atom is known, the atomic weights of all the other elements may be found immediately from the chemically determined *relative* atomic weights. Thus the weight of the oxygen atom is 16 times, and the weight of the lead atom is 207 times, the hydrogen value.

That the ultimate particles of matter, be they recognized as atoms or as molecules, are in *perpetual motion* has been a common idea associated with atomism throughout its history. In the seventeenth century, as stated before, heat was identified with the energy of this motion; and in 1851 James Prescott Joule (1818-1889), an English brewer and amateur scientist, estimated the speed with which the molecules move. He did this by deducing theoretical formulas relating the observed gas pressure to the average impulse of molecular collisions with the walls of the container.

The molecular speeds determined by Joule were surprisingly high. At ordinary temperatures the molecules of air fly about at an average speed of four hundred meters per second (a meter is a bit more than thirty-nine inches), which is the speed of rifle bullets, and at higher temperatures they move even faster.

Scientists, particularly physicists, are greatly impressed by numerical values. Thus, the increasing store of quantitative data about atoms and molecules served to enhance the sense of the reality of these minute particles.

DISCOVERY OF THE ELECTRON

THE next step forward in the study of atoms came from experiments in electricity. Although material substance is normally devoid of electrical effects, ways of producing such effects were known to Greek scientists several centuries before the Christian era. They found that, when amber (the Greek word for it is *ēlektron*, from which "electricity" is derived) is rubbed with a cloth, it acquires the ability to attract bits of straw and other light materials. This remained a curious, isolated fact until well into the eighteenth century. Newton, Boyle and other seventeenth-century scientists had observed electrical effects. But aside from showing that glass and various other substances display the same effects as amber, they contributed nothing new.

The eighteenth century brought an upsurge of interest in electrical experiments. Various devices were designed to produce more pronounced electrical effects; and demonstrations with vigorous electric sparks, with their attendant noise and flashes and "sulphurous smells," came to be much in vogue in high society and in popular lectures.

More serious experiments disclosed that there are two kinds of "electric fluid," the *vitreous*, obtained by rubbing glass, and the *resinous*, the kind the ancients had found by rubbing amber. Of great significance was the further discovery that two bodies bearing the *same* kind of electric fluid (both vitreous or both resinous) *repel* each other by an electrical force while two bodies having *different* kinds (one vitreous and the other resinous) show a mutual *attraction*.

Benjamin Franklin (1706-90), American printer, publisher, author, inventor, diplomat, statesman, scientist and philosopher, devoted a goodly portion of his tremendous energy to the study of electrical effects. It was he who proved that lightning is a huge electric spark by drawing down the "electric fire" from thunderclouds along the string of a kite—an ingenious and bold experiment in which he had the great good luck to avoid electrocuting himself. (Another who tried the experiment later at St. Petersburg, in Russia, was not so lucky.)

Franklin proposed a simpler, one-fluid theory according to which bodies having more than the normal amount of the electrical fluid are described as being charged *positively* while those having a deficiency are *negatively* charged—positive corresponding to vitreous charge. This theory explained all electrical phenomena known in Franklin's day. Later investigations into the nature of electricity showed that there actually are two kinds of electric charge, for which Franklin's terminology of positive and negative (or *plus* and *minus*) was retained.

The law of electrical force was discovered by the French physicist Charles Augustin de Coulomb (1736-1806). He showed that the force between two small electrically charged spheres depends directly on the product of the charges and varies inversely as the square of the distance between their centers, another "inverse square" law similar to Newton's law of universal gravitation. The electrical force is, however, far stronger and is easily observable; and it may be either an attraction or a repulsion while gravity is always an attractive force (and becomes appreciable only if one of the bodies, that is, the earth, has a very large mass).

The idea that electricity might be atomistic, that it might consist of tiny, discrete particles of electric charge, had been proposed repeatedly throughout the nineteenth century. When it was found, in the 1870's, that a current of electricity could be made to flow through a highly evacuated glass tube, it seemed that this current, through nearly empty space, must be a stream of pure electricity, existing quite apart from ordinary matter. It ought, therefore, to present a unique opportunity to discover the nature of electricity itself.

This problem was attacked with great skill by the English physicist Sir Joseph John Thomson (1856-1940). By subjecting the stream of electricity to the action of electric and magnetic forces, he showed that it does indeed consist of discrete particles; and he also succeeded in obtaining fairly good values for their individual electrical charge and mass. The charge was found to be *negative* (in Franklin's scheme) and equal to 1.6×10^{-9} coulomb, a coulomb being approximately the quantity of charge that flows through a 100-watt electric light bulb in one second. But, of more significance for the story of atomic structure, the mass was found to be only about 10^{-27} gram, roughly two thousand times less than that of the lightest atom.

These ultimate particles of electric charge, later called *electrons* (and designated by the letter e) were found to be ubiquitous. Whenever atoms of any kind were subjected to rough treatment, as when buffeted about in hot flames or electric sparks, electrons appeared. It became evident that these particles are a common constituent of all atoms, that they are an elementary "building block" of the material universe. This in itself was a momentous discovery. But it carried an implication of even greater moment: if atoms have constituent parts, they can no longer be thought of as ultimate entities. Thus, the hard, indestructible, indivisible atoms of Democritus ceased to exist.

Thomson received the Nobel Prize and was later knighted for

his scientific achievements. In his post as director of the Cavendish Laboratory at Cambridge University, he made England pre-
eminent in experimental physics during the first three decades
of the twentieth century. His outstanding capability as leader
and teacher is attested by the fact that seven of the young men
who worked under his guidance later received the Nobel Prize.

THE ATOM MODELS OF THOMSON AND RUTHERFORD

BECAUSE atoms in their normal condition are electrically
neutral, it follows that they must have, in addition to their com-
plement of negative electrons, an equal amount of *positive* charge
so as to make their charge, as a whole, equal to zero. Since no bits
of positive charge had ever been separated from atoms, it seemed
that the positive part must be all of one piece. The mass of the
electrons having been found to be negligibly small compared to
that of atoms, it followed further that the positive part must
represent nearly all of the atom's mass and should presumably
form most of the bulk.

From all of this evidence Thomson fashioned in 1904 a *model*
of the atom, which is noteworthy as the first serious attempt at a
detailed quantitative concept. (A model in the sense used here
is not a small replica, as a toy model airplane, but an imagined
construct to account for observed phenomena.) He pictured the
atom as a small sphere of positively charged stuff with much
smaller negative electrons embedded in its "like plums in a pud-
ding," the plums being arranged, however, in neat circles.

There was no direct experimental evidence in support of Thomson's atom model, and the task of obtaining such evidence must
have seemed well-nigh impossible. It did require the skill and
imagination of an experimenter of the highest caliber, none
other than one of Thomson's distinguished seven, Ernest Rutherford (1871-1937).

Rutherford was born on his father's farm near the small town of
Nelson in New Zealand. He showed great ability in his studies and
was therefore sent to the University of New Zealand, where he received a prize scholarship to study at Cambridge—the first man from

"down under" to be so honored. He worked at the Cavendish Laboratory under Thompson's guidance and later for several years at McGill University, in Canada. In 1907 he was called back to England to direct research at the University of Manchester, where his work was so outstanding that he was made director of the Cavendish after Thomson retired in 1918 to an honorary professorship. He was president of the Royal Society from 1925 to 1930 and was given the title of Baron Rutherford of Nelson in 1931.

At McGill, Rutherford had started research on radioactivity, that strange, new phenomenon investigated by Pierre (1859-1906) and Marie Curie (1867-1934), whose lives were a poignant drama of dedication and sacrifice. He named the two kinds of radiation given off by radium and other radioactive elements *alpha* and *beta* (to which a third, *gamma*, was added later). The beta radiation was soon found to consist of Thomson's newly discovered electrons. But the nature of the alpha radiation was more obscure, and Rutherford set himself the task of finding out all he could about it. He soon discovered that the alpha rays are positively charged particles, the positive part of helium atoms. He found that they can penetrate through thin layers of solids, for example, into the interior of a thin-walled glass tube without making any perforations, presumably going right through the atoms themselves. He decided that it might be possible to investigate the internal construction of atoms by shooting these tiny bullets through them.

To test this idea, Rutherford performed an experiment with the assistance of Hans Geiger (1882-1945) who had come from Germany to work in his laboratory. Alpha particles, coming from a bit of radium, were made to pass through a thin gold foil and onto a *scintillation screen*, a glass plate covered with a layer of zinc sulfide, which emits a tiny flash of light where an alpha particle strikes (just as the face of a television tube lights up where it is struck by a stream of high-speed electrons). The expected happened; the alpha particles passed through the foil and nearly straight ahead to the screen.

In some earlier experiments Rutherford had, however, found indications that a very few alpha particles might be deflected more widely while passing through atoms, and he suggested that this possibility ought to be looked into more carefully. Accord-

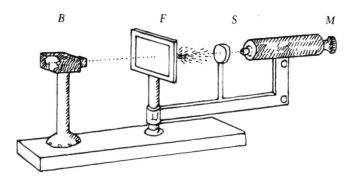

Figure 2. Rutherford's alpha-particle bombardment apparatus.

ingly, Geiger built an apparatus whose principle of operation is shown in Figure 2. (In the actual apparatus the space through which the alpha particles pass was evacuated to avoid the disturbing effects of collisions with air molecules.) A bit of radium enclosed in a lead box *B* sends out a narrow stream of alpha particles which pass through the thin gold foil *F* and are scattered off at various angles onto the fluorescent screen *S*, where the flashes produced by each impact are viewed with the microscope *M*. The screen and microscope are mounted on a pivot so that they may be moved on a circle around the foil to catch particles coming off at any angle.

Again almost all of the particles went nearly straight through the foil. But very rarely, once in several thousand, a particle veered far off to the side, even straight back toward the source. Rutherford described this result as "quite the most incredible that had ever happened to me in my life." Simple calculations showed that neither the electrons nor the positive part of the Thomson atom could exert the required strong sideward push on the speeding alpha particles.

This unexpected result gave the experiments greatly increased importance. Geiger and his assistants spent many weary hours sitting in a darkened room, straining their eyes to catch the rare deflections coming off at large angles. Great numbers of observations were made, using foils of various other metals in addition to gold.

As the data were accumulating, Rutherford worked out a mathematical analysis based on the conception of the atom shown in Figure 3. He assumed that the positive part of the atom is not spread throughout the whole of it but is concentrated in a tiny *nucleus*, much smaller than the whole atom but still, astonishingly, containing near all of its mass. An alpha particle heading for a point close to this nucleus would be given a strong push to one side and deflected sharply as shown; one passing through the atom at a greater distance from the nucleus would be deflected only slightly. Assuming that the deflecting repulsion between alpha particle and nucleus is Coulomb's inverse square force, he obtained precise numerical agreement between his calculations and the results of the experiments.

Rutherford's calculations showed that the size of the nucleus must be less than 10^{-12} centimeter, one-ten-thousandth the size of the whole atom. (In the figure the electrons and the nucleus are shown far too large; if drawn to proper scale, they would be too small to be visible.) Further, he found that the charge on the

Figure 3. The Rutherford planetary atom model showing the paths of alpha particles passing the nucleus.

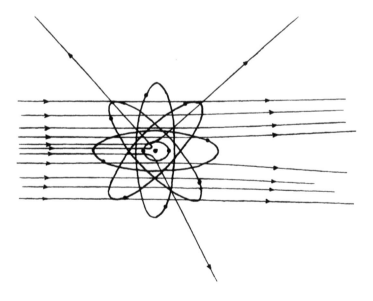

atomic nuclei of the various metals tested is equal in magnitude (though opposite in sign) to the electronic charge e multiplied by the atomic number of the metal. This number, usually written Z, must therefore also signify the number of electrons in the atom. The atom of gold, for example, whose atomic number is 79, has a nucleus carrying a charge of $+79e$ surrounded by 79 electrons, each with a charge of $-e$. This fact had been suggested by other observations. But Rutherford's alpha particle bombardment experiments, together with his mathematical analysis, afforded direct and convincing evidence for this important interpretation of atomic number. (To Mendeleev the atomic number had been merely a way of indicating the sequence of elements in his table, with no more intrinsic significance than the numbers on theater seats.)

The electric charge of the hydrogen nucleus is just $+e$; and this smallest positively charged particle, which Rutherford named the *proton*, was recognized, along with the electron, as a second elementary constituent of matter. For a time (up to 1932) it seemed that all atoms, and thus the entire physical universe, might be constructed of just these two kinds of particles, a tremendous simplification compared to the ninety-two distinct kinds of atoms in the chemical atomic table, and by far the simplest conception of the structure of matter ever devised by the mind of man.

According to this model, the atom is nearly all empty space since the particles composing it are extremely minute compared to the size of the whole ("like a few flies in a cathedral," as Rutherford put it). It follows that solid bodies are also nearly all empty space. They appear solid because the group of electrons in each atom resists intrusion by the electrons of adjacent atoms; each atom thus appropriates its tiny region of space unto itself much as if it were a minute solid particle.

The Rutherford atom model, based on substantial experimental evidence, seemed to merit consideration as a possible description of actual atomic structure; and it was reasonable to demand that it comply with the laws of classical physics. But here a difficulty arose with regard to its *stability*, for two objects carrying electrical charges of opposite sign attract one another. Therefore, the positive nucleus should quickly draw the negative electrons in upon itself, thus causing the atom to collapse. Rutherford suggested that the electrons might be traveling in circular orbits about the nucleus much as planets move around a sun and keep from falling inward in spite of the gravitational attraction acting upon them. But this idea posed a new difficulty, for according to the classical laws of electrodynamics, a circulating electron must necessarily emit a ray of light, thereby dissipating its energy and spiraling in upon the nucleus. Yet atoms, and the universe constituted of them, obviously continue to exist.

It might seem surprising that scientists, supposedly sensible and knowledgeable, should give such an apparently impossible contrivance as this *planetary* atom model even a moment's consideration. The explanation lies in their appreciation of the distinction between fact and fancy. Facts, obtained from observation and measurement, are treated with the greatest respect; any theory, be it in the form of an imagined model or a mathematical formula, which contradicts experimental observations is immediately suspect. The Rutherford model did account for the scattering of alpha particles; but this effect depended only upon the existence of the tiny, massive nucleus. The rest of the picture, the electrons whirling about in their orbits, had *not* been related to any experimental evidence. Therefore, although its imagined characteristics were not in agreement with the laws of classical physics, it could be retained as a tentative hypothesis, open to further experimental investigation.

Then, in the hands of Niels Bohr, the Rutherford atom model was indeed shown to account for an imposing array of experimental observations. Some of the more important of these are concerned with the interaction of atoms and light rays, in which atoms exhibit their *dynamic properties.* These will be considered later after the nature of light has been discussed. First it is necessary to present the new ideas which Bohr developed concerning the *structure of atoms.*

THE BOHR ATOM MODEL

NIELS HENRIK DAVID BOHR (1885-1962), whose distinguished career spanned the entire development of atomic physics, was the son of a professor of physiology at the University of Copenhagen. He received his Ph.D. there in physics and, in 1911, was awarded a scholarship for study abroad. Like many other young physicists of his day, he traveled to the renowned Cavendish Laboratory at Cambridge to study under J. J. Thomson. His sure intuition led him to divine that Rutherford's ideas about atomic structure probably lay closer to reality than did the Thomson model. Since "J. J." thought rather well of his own conception and was somewhat sensitive to criticism of it, Bohr soon repaired to Manchester to join Rutherford's group.

With that clear grasp of reality which was to mark all of his work, Bohr fixed his attention on what actually appears to happen rather than what classical laws demand. He assumed that, for reasons as yet unknown, the electrons in Rutherford's atom model can continue to circulate about the nucleus in fixed orbits without radiating any energy in the form of light. The radical new ideas he developed concerning the way in which atoms *do* radiate light are part of the story of how atoms behave when they are jogged into special activity. For the moment our concern is with atoms in their normal state.

Concerning the *nucleus* of atoms, Bohr adopted the findings of Rutherford and others to the effect that its electrical charge is positive and is equal, in electronic charge units, to the atomic number Z. The simplest assumption to account for this is that all nuclei are closely packed groups of protons. But this fails at the very first step beyond hydrogen. While the helium nucleus has a charge of two protons, it has the mass of four. This could be explained on the assumption that this nucleus consists of four protons, to make up the proper mass, plus two electrons (whose contribution to the mass would be negligible) to neutralize the charges of two protons.

This scheme was actually used to account for the electrical charge and the mass of all the various atomic nuclei. But, to anticipate later events, the picture of nuclear structure was simplified by the discovery in 1932 of a third kind of elementary particle, the *neutron*, having a mass nearly equal to that of the proton (the proton has a mass 1836 times greater than the electron; for the neutron the value is 1839) but, as the name implies, it has no net electrical charge. The helium nucleus could then be depicted as a very tightly knit cluster of two protons and two neutrons. Similarly, the nucleus of the oxygen atom, with an atomic number 8 and an *atomic weight* W of 16, is made up of 8 protons and 8 neutrons. In the copper atom, for which Z is 29 and W is 63, the nucleus contains 29 protons and 34 neutrons. Finally, the nucleus of uranium with 92 protons has, in addition, 146 neutrons to make up its atomic weight of 238. (There are numerous examples of two kinds of atoms of the same chemical element whose nuclei have the *same* number of protons but a *different* number of neutrons. For example, there are two kinds of chlorine, both with 17 protons but with, respectively, 18 and 20 neutrons. Nearly all the elements have been found to have two or more such variants, called *isotopes*).

With this new concept of nuclear structure there remained the important question of how the component parts of the nucleus hold together against the mutual repulsion of the positively charged protons. No answer to this question was at hand; it had to await the development of new scientific concepts which showed that a binding force acts among protons and neutrons which is quite distinct from the previously known gravitational and electrical forces.

The process of building up the sequence of ninety-two different kinds of atoms, starting from hydrogen with its single proton and single orbital electron, could be depicted as adding at each step one proton and a number of neutrons to the nucleus and one additional electron to the group circulating around it. If these electrons were added one by one to a steadily increasing group, the atoms in the sequence would presumably grow steadily in size (but not greatly, since the increasing nuclear charge would exert an increasing contractive force on all the orbits), and, in general, the sequence of atoms would exhibit a *gradual* trend in all of their various properties.

The observed facts are utterly different. Proceeding through the table of elements, many properties show a *periodic* mode of variation. Atomic sizes, for example, repeatedly increase and then decrease again with rising atomic number. Most importantly (and this is why Mendeleev's table is called periodic), similar *chemical* properties appear repeatedly at various points in the sequence. Thus, the *alkalies,* such as lithium, sodium, and potassium (appearing in column I of the periodic table, page 18), all highly caustic, stand at atomic numbers 3, 11, 19, and so on; and similarly, four of the *halogens*, fluorine, chlorine, bromine, and iodine (in column VII), all strongly acid-forming, stand at numbers 9, 17, 35 and 53.

All this, it seemed to Bohr, indicated strongly that, corresponding to this periodicity in their observed *properties*, there must be some basic periodicity in the sequential *structure* of the atoms. This, he decided, could come about through an arrangement of the electron orbits in groups or *shells*, one outside the other. As each shell is completed, the start of a new shell occurs over and over again. While it seemed evident that each successively larger shell should be able to accommodate more electrons, much patient sifting of chemical data was required to determine the following scheme for the *maximum* number which successive shells can hold:

Shell number	1	2	3	4	5	...
Numbers of electrons	2	8	18	32	50	...

These shells are also commonly designed at as *K, L, M, N, O,* and so forth.

In building the sequence of atoms, the first two electrons are placed in shell number 1; the third, at *lithium*, starts shell number 2. After this is completed, the eleventh electron, at *sodium*, starts number 3. It is to be expected that the chemical characteristics of an atom, the way in which it interacts with other atoms, should depend largely on the number of electrons in the *outermost* shell, since a chemical interaction is a meeting of atoms at their periphery. The shell structure thus accounts nicely for the chemical similarity of lithium and sodium since both have a single electron in the outer shell. With the atoms of higher atomic number the interpretation of chemical data is complicated by the fact that a new shell is started before the previous one is completed. Electrons are added to shell number 4, for example, after number 3 is partially filled with 8 electrons; and similar complications occur in the larger shells, as is obvious in the periodic table of elements, page 18.

It is found that an outer shell of 8 electrons is apparently a preferred configuration which atoms "strive" to attain in their chemical interactions. This is exemplified by the reaction in which

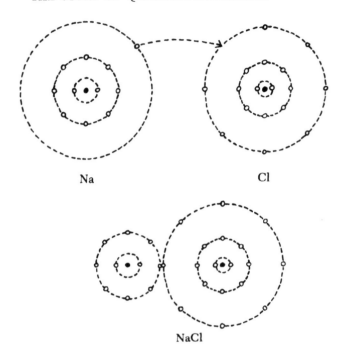

Na   Cl

NaCl

Figure 4. The formation of a sodium chloride molecule.

sodium and chlorine unite to form sodium chloride, common table salt. The process is shown in Figure 4, in which the atoms are depicted in a highly stylized manner with electrons in flat rings symbolizing the shells. Sodium (Na stands for its Latin name, *natrium*), as has been explained, has a single electron in its third shell while chlorine (Cl), with 17 electrons, has 7 in this shell. By a transfer of the single electron from the sodium to the chlorine atom, both acquire the 8-electron configuration. This transfer leaves Na with a net positive charge of $+e$ while that of Cl has become $-e$; and since unlike electrical charges attract, the two are bound together as a molecule of NaCl. This explains why Na, a strongly corrosive caustic, and Cl, a highly poisonous gas, combine to form harmless table salt. In the process both have lost that electron configuration which gives them their characteristic chemical properties.

All chemical reactions are not explained in such a simple manner. Nevertheless, Bohr's scheme of the shell structure of atoms accounts successfully for a vast array of chemical data. To him must go the credit for having provided an explanation for the periodic table of elements. The idea of a preferred configuration of 8 external electrons had been proposed earlier by two American chemists, G. N. Lewis (1875-1946) and Irving Langmuir (1881-1957). They pictured these electrons as being arranged at the eight corners of a cube, an idea which accounted for the chemical behavior of many kinds of atoms.

The cryptic numerology involved in the sequence of numbers designating the maximum number of electrons in successive shells (dividing them by two gives the numbers 1, 4, 9, 16, 25 . . . , which are the squares of the successive natural numbers 1, 2, 3, 4, 5 . . .) was highly intriguing. It could not possibly be a mere happenstance, and yet there was no hint of an explanation for it. As it stood, the whole Bohr scheme was entirely empirical, derived from experimental findings, though guided by a superb intuition. Many new and unforeseeable insights were to be achieved before these numerical relations came to be understood, and much of the detailed content of Bohr's picture was to be swept aside by future events. But the relations to the natural numbers, which held the essence of its worth, remained unchanged.

In working out the details of the electron orbits and shells, much help was afforded by spectroscopic data, by observations on the interaction of atoms and light. Moreover, these were primarily the observations that prompted Bohr to take up the study of atoms. It is appropriate therefore to consider the developments in optics, notably those of the early twentieth century, which he put to excellent use in formulating his ideas about the dynamics of atoms.

4

Radiant Energy

LIGHT WAVES

THE physical universe which has been presented thus far is built of three kinds of elementary particles: electrons, protons and neutrons. These, in turn, are made of *mass* and *electrical charge,* the two fundamental entities constituting all material substance. It is now necessary to add a third which, at the present stage of our story, stands apart as a new and distinct entity: *radiant energy.* The light and radiant heat of the sun, which streams down upon the earth through ninety-three million miles of empty space, obviously has its existence quite apart from matter. (Later it will be seen that the two are related in certain ways.)

There is much experimental evidence to show that radiant energy has wavelike properties. A wave, as seen in Figure 5, is characterized by its *amplitude* (*A*), or the height of the crests;

Figure 5. The characteristics of a sinusoidal wave.

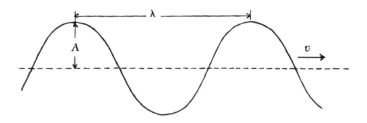

its *wavelength* λ (lambda), or the distance from crest to crest; and its *frequency* ν (nu), which is the number of undulations per second, or the number of crests that pass a given point per second, as the wave travels by. It is further characterized by the velocity v at which it moves ahead. The significance of these characteristics will be made evident presently.

Radiant energy exists in many forms which differ from one another in their wavelengths. The range of wavelengths is enormous, as shown in Table 2, where the various kinds are listed in order of decreasing wavelength. The longest are the radio waves, which carry speech and music from radio sending stations to our homes. The shortest are the X-rays. Gamma rays are identical with X-rays but are produced naturally by radium and other radioactive atoms.

It is evident that the wavelength range of radiation that is visible to our eyes, the *visible spectrum of light*, includes but a minute portion of the whole tremendous scope of wavelengths. The longest of these visible waves appear to us as red light; with decreasing wavelength the colors range through orange, yellow, green, blue, indigo and violet. The infrared and ultraviolet regions lies beyond the long and short wave limits of the visible spectrum.

Ordinary white light, from the sun and other luminous sources, is a mixture of all these colors. The yellow glow of candles is richer in the longer wavelengths while the dazzling blue-white light of an electric arc contains more of the shorter blues and violets.

All these waves, regardless of wavelength, travel through empty space at the same velocity—the *velocity of light*, which is 186,000 miles per second or three hundred million (3×10^8) meters per

Table 2. Wavelength ranges for the various kinds of radiant energy.

Type of radiant energy	Wavelength range
Radio waves	10,000 to 1 meter
Radar waves	10 to 0.1 centimeter
Radiant heat	10^{-1} to 10^{-3} centimeter
Infrared	10^{-3} to 10^{-4} centimeter
Visible light	8×10^{-5} to 4×10^{-5} centimeter
Ultraviolet	4×10^{-5} to 10^{-7} centimeter
X-rays and gamma rays	10^{-8} to 10^{-12} centimeter

second. This velocity, usually designated as c, is a universal constant of nature that plays an important role in the theories of relativity.

For anything that lopes along in jumps, the velocity of travel is equal to the length of each jump multiplied by the number of jumps per second. Thus, a rabbit making ten-foot jumps at the rate of three per second is moving along at thirty feet per second. Similarly, the velocity at which a wave travels is the product of its *wavelength* and its *frequency*. Radio waves, with a "stride" of three hundred meters, step along at the rate of a million "steps" or undulations per second; and the frequency at which the minuscule light waves must step, to keep abreast of the giant strides of the radio waves, well-nigh staggers the imagination. It is indeed a large number, about 10^{15} undulations per second. It is evident that shorter wavelengths mean higher frequencies while lower frequencies imply longer wavelengths.

With the usual symbols all of the above may be expressed briefly in the simple relations:

$$c = \lambda \nu \qquad \lambda = c/\nu \qquad \nu = c/\lambda$$

The concept that light is a wavelike phenomenon brought up a serious difficulty. While waves on a lake are obviously occurring in the water, and the sound waves coming to our ears are less obviously carried by the air, scientists were confronted with the problem of explaining how light waves would be propagated through space that is completely devoid of material substance. They attempted to cope with this difficulty by inventing the "luminiferous ether," an utterly mysterious, invisible, impalpable "subtile fluid" that was supposed to pervade all space and all objects in it.

Our century has seen the existence of an "ether" challenged by the theory of relativity and by ingenious experiments designed to test that theory. But these matters lie outside the province of our interest, and for the present the concept of light waves may be accepted without concern for the question of what is waving. Later insight regarding the nature of light (which will be discussed in considerable detail) have made this issue largely irrelevant.

Nineteenth-century physicists were thoroughly convinced of the wavelike nature of light. But this had not been the case in the two preceding centuries. Indeed, right up to the end of the eighteenth century the prevalent opinion had been that light consists of tiny speeding corpuscles. One of the chief arguments in favor of the corpuscular versus the wave theory of light was based on its observed *rectilinear propagation*.

That light travels in straight lines is a common observation. It is possible to see where things are because the light coming from them travels to our eyes in straight lines. Surveyors make use of this fact when laying a roadway, and astronomers depend on it to determine the precise location of the stars. Waves are known, however, to wash around corners and thus to depart from straight lines. A person talking in an adjacent room can be heard because the sound waves he emits bend around the door frame to reach our ears; but unless he is in the direct line of sight, he cannot be seen since light does not turn the corner. Or so it seems.

It is quite easily demonstrated, however, that under proper circumstances light does actually bend away from straight lines of travel. All that is needed for the experiment is a small source of light. A candle flame or a lighted flashlight seen across a room or a street lamp down the street at night will do. Holding up two fingers vertically close to one eye, the light source is observed through the space between them as they are gradually brought closer together. Just as this space becomes very narrow, just before the fingers touch, something remarkable is observed. The light, which up to this point looked perfectly normal, is suddenly seen to spread out into a wide, horizontal band of light.

What is seen here is due to a fundamental property of all waves, of water waves and sound waves, and therefore also of light if it has wavelike properties. The effect is evident in Figure 6 showing photographs of ripples on the surface of water. Passing through an opening that is considerably greater then their own wavelength, as in (1), the ripples proceed nearly straight ahead, but after passing through a narrower opening at (2), they spread out widely. Similarly, when light encounters a wide slit, as shown at (1) of Figure 7 (which indicates a single beam of light from a candle), it passes straight through; only if the slit is extremely narrow does it spread out beyond. At (2) of the diagram, the light from a small source S passes through the narrow slit and spreads out into a bundle of rays that enter the eye in a range of angles. Since things appear to be in the direction from which the

(1)

(2)

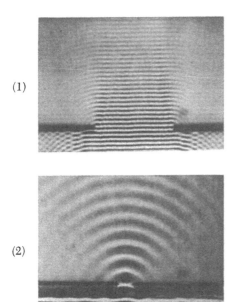

Figure 6. Photographs of ripples on the surface of water.

(3)

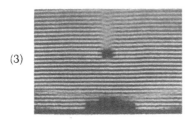

rays reach the eye, the light source seems to extend in a streak from S_1 to S_2.

Quite generally, when waves encounter obstacles of any sort that are large compared to their own wavelength, they pass on very nearly straight ahead. It is when they meet things that are comparable in size to their wavelength that they wash around them and depart conspicuously from rectilinear propagation, as seen in photograph (3) of ripples passing a small object. The effect depends on *relative* size. Ripples on a pond will wash around a reed protruding through the surface of the water, but will proceed in straight paths beyond a large rock, leaving a "shadow," free of ripples, behind it. Ocean waves wash around the rock, but there will be a region devoid of waves behind a large breakwater built to protect a harbor.

Those who argued against a wave theory of light did not realize that the wavelengths of visible light are only hundredths of a hair's breadth, exceedingly minute compared to those of sound. The latter are of arm's length, and for them a door is quite narrow. It is obvious, therefore, why someone in an adjacent room can be heard through the doorway though he cannot be seen if he is out of the direct line of sight.

Diffraction is the name given to the bending of light into the shadows, and diffraction plays an important role in many natural phenomena. It is a simple matter, for example, to construct a microscope with which one can see objects a hundred times smaller than those visible to the unaided eye. It would seem that unlimited magnification could be obtained by arranging a number of microscopes one above the other so that each one views the image of the one below and magnifies it again a hundredfold. But diffraction sets a limit to this. It is not possible to obtain a clear image of any object that is as small as the wave-

Figure 7. Paths of light rays through wide and narrow slits.

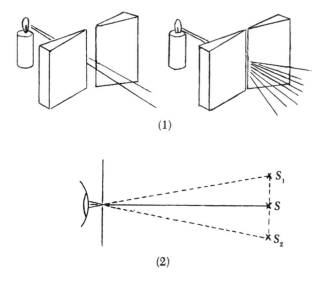

(1)

(2)

length of light since the waves passing it do not proceed in straight paths to the eye.

INTERFERENCE OF LIGHT WAVES

EARLY in the nineteenth century a simple and ingenious experiment was devised that provides convincing evidence for the wavelike character of light. It was performed by the English scientist Thomas Young (1773-1829), an infant prodigy who read the books in his father's library fluently at the age of four and grew up to be a versatile scientist with various important contributions in physics and physiology to his credit. The essential parts of his apparatus are shown schematically in Figure 8. L is a source of light; S_1, S_2 are a pair of very narrow, parallel slits cut in an opaque sheet. Light streams from the source through the slits and falls upon a screen set a foot or two beyond them. The slits are actually less than a hair's breadth in width so that the light, after passing through them, spreads out broadly, light from both slits falling over a wide expanse of the screen.

As his source, Young used sunlight coming through a small hole into a darkened room. But to make matters simple, it will be assumed that a sodium vapor lamp is used since it emits but a single color of yellow light and therefore only a single wavelength. With this apparatus he observed on the screen a pattern of bright (B) and dark (D) bands parallel to the length of the slits, and he succeeded in formulating a simple and convincing explanation for this light pattern.

The essential concept in his account is the *interference* of light waves. He assumed that, if two light waves meet *in phase* (both waving up and down together), they reinforce one another, as shown, to produce a strong resulting wave, while two that meet a half-wave out of phase (one waving up, the other down) destroy one another and produce darkness.

Since both slits get their light from the same source, the waves starting from S_1 are in phase with those from S_2. (This is an essential condition; using separate sources of light in place of the two slits would not do.) The waves traveling along the two paths of equal length to the center of the screen at B_1 arrive in phase and produce a bright band. At a point D_1, a short distance off to the

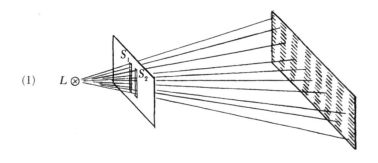

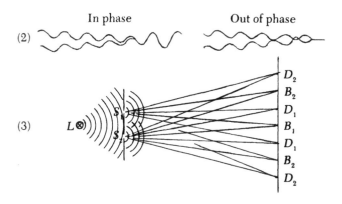

Figure 8. Young's double slit apparatus demonstrating interference of light waves. (1) The general arrangement of light source, slits and screen. (2) The constructive and destructive interference of two waves. (3) The production of a sequence of bright and dark bands by interference of light rays from both slits.

side, where the two paths from S_1 and S_2 differ in length by just a half wavelength, the waves meet out of phase and produce a dark band. Further to the side, at B_2, the paths differ by a whole wavelength, the two waves are effectively in phase again, and a second bright band results. The pattern thus goes on in alternate bands of light and dark, and it is repeated on the other side of the central band. By measuring the spacing of these bands, it was possible to obtain an approximate value for the exceedingly short wavelengths of light, an outstanding achievement for experimental physics.

It is obvious that the corpuscular theory of light could not account at all for these observed effects. If light behaved like tiny bullets, it would pass from the source through the slits and straight ahead to two lines at either side of the band pattern on the screen. In particular, no light at all would reach the central band, which is actually observed to be the brightest part of the light pattern. And most emphatically, corpuscles spraying onto the screen from both slits could not posibly add up at any point to no corpuscles.

In England, where the authority of Newton, who had favored a corpuscular theory of light, was still strong, Young's findings were not quickly recognized. It remained for a group of French scientists under the leadership of Augustin Jean Freznel (1788-1827) to carry out further investigations. Freznel devised a number of ingenious experiments to demonstrate the wave nature of light and developed an effective theory of optics based on the wave concept. His work stimulated widespread interest in optical experimentation, all of which provided additional evidence in favor of waves. Then, in 1870, the English theoretical physicist James Clerk Maxwell showed that light is intimately related to electrical and magnetic phenomena and developed the *electromagnetic wave theory* of light, which was soon given experimental support through the work of the German physicist Heinrich Rudolph Hertz (1857-94). The corpuscular theory was quite forgotten, and toward the close of the nineteenth century there was a complete consensus among physicists that light is a wave phenomenon.

LIGHT PARTICLES

IN 1900 the German theoretical physicist Max Planck (1858-1947) published a paper on radiant energy wherein he set forth a new idea, completely foreign to classical science, which was destined to become one of the salient conceptions in the physics of the twentieth century. It is remarkable that this decisive break with classical physics was made by a man who was steeped in the purest classical tradition, as it was nurtured in the German universities toward the close of the nineteenth century. Planck lived to the age of ninety; he survived both world wars and was able to witness the destruction of Nazism, which he detested. In 1918 he was honored with the Nobel Prize for his original ideas, and it was actually for work following up these ideas that both Niels Bohr and Albert Einstein (1879-1955) were awarded the Nobel Prize some years later.

To understand the problem which Planck strove to solve, it is necessary to go back a decade before Rutherford's experiments, to the time when no serious attempts had yet been made to formulate actual atom models. Atoms were still thought of vaguely as tiny structures containing electrically charged particles held by springlike fastenings that would permit them to tremble or oscillate when the atoms receive a jolt. It was also known that the atoms in solid bodies are in constant, random, to-and-fro motion and that this motion becomes more violent as the bodies become hotter. The mutual joltings would set the electrified particles within the atoms, the *electronic oscillators*, into motion, causing them to emit waves of light and radiant heat, somewhat as a bobbing cork sends out ripples on the surface of a pond. This was the mechanism to account for the light and heat radiated, for example, from the glowing coals of a fireplace.

Since it was known that, as bodies get hotter, they send out progressively radiant heat, infrared, visible light (changing from dull red to brilliant blue-white), and when extremely hot even some ultraviolet radiation, it was necessary to assume that within the myriad atoms of the body there are numerous electronic oscillators capable of vibrating at all the frequencies corresponding to this whole range of wavelengths.

In spite of the vagueness of the assumed mechanism of radiation, the classical laws of mechanics and electrodynamics could make very definite predictions about the kind of radiant energy a hot body should emit. These laws state that the energy supplied to the radiating oscillators will be apportioned equally among them (the *equipartition theorem*) and also that those of higher frequency (and shorter wavelength) will radiate their energy more effectively. Accordingly, bodies at all temperatures should radiate more energy in the violet and ultraviolet than in the infrared and red; and even moderately hot bodies should emit light of the same intense blue-white color, though of lesser amount, as do those which are at a dazzling, white-hot temperature. This completely erroneous prediction of classical laws was dubbed the "ultraviolet catastrophe." (If X-rays had been known at the time, the situation would have been more catastrophic.)

This outcome was the cause of much concern; it was indeed one of the darkest "clouds" on the horizon of classical physics. The physicists Lord Rayleigh (1842-1919) in England and Wilhelm Wien (1864-1928) in Germany, among others, worked on the problem without success. Planck, however, found an approach which gave a theoretical result in complete agreement with experiment. But to get this result he had to make a radical and seemingly absurd assumption, for according to classical laws, and common sense as well, it had been presumed that an electronic oscillator, once set in motion by a jolt, radiates its energy smoothly and gradually while its oscillatory motion subsides to rest. Planck had to assume that the oscillator ejects its radiation in sudden spurts, dropping to lesser amplitudes of oscillation with each spurt. He had to postulate that the energy of motion of each oscillator can neither build up nor subside smoothly and gradually but may change only in sudden jumps. In a situation where energy is being transferred to and fro between the oscillators and the light waves, the oscillators must not only *emit* but also *absorb* radiant energy in discrete "packets." At any given moment, therefore, any one oscillator contains either one or two or any whole number of such packets of energy. Any content of energy between these discrete values is, for some unknown reason, excluded.

The essential assumption in Planck's scheme concerned the *size* of energy packets. He postulated that the discrete quantities of energy which an oscillator can absorb or emit are directly proportional to its *frequency*. Oscillators of low frequency can accept energy in small packets; those of high frequency will accept only large ones. This immediately does away with the ultraviolet catastrophe. For in a body at moderate temperature the joltings of the atoms are sufficiently energetic to supply the requirements of the low-frequency but not of the high-frequency oscillators, and these do not receive any energy to radiate. Only those of lower frequency are active, and the body radiates nothing but the long wavelengths. With increasing temperature, the atomic jolts become more energetic, oscillators of higher frequency are activated, and the character of the radiation shifts toward shorter wavelengths.

Planck obtained not only this general agreement with observations; when he worked out the mathematical consequences of his
postulates, they were in precise agreement with the very careful
measurements of the experimenters. He coined the name "quanta"
for the packets of energy, and he spoke of the oscillators as being
"quantized." Thus, the trenchant concept of the *quantum* entered physical science.

In spite of its impressive triumph, physicists were slow in ac-
cepting Planck's radical conception. He himself disliked it and
hoped that someone would find an adequate explanation for the
radiation of hot bodies more compatible with classical principles,
and that the quantum would go away. But on the contrary, the idea
of energy quantization grew steadily in importance and became
the dominant concept in the further development of atomic
physics.

It was found that all *periodic* motions, which go to and fro or
round and round repeatedly, have their energy quantized. Such
a common occurrence, it would seem, should be easily observable;
and it is appropriate to ask why it is not evident, for example, in
the swing of the pendulum. Apparently, by nudging a pendulum
very gently, its energy of swing may be built up gradually; and
when left alone, its oscillation subsides smoothly again. Indeed,
it would be surprising to see the oscillation decreasing in a series
of sudden jumps. In principle, the energy, however, does change
in jumps. But for a large-scale pendulum the individual jumps
are too small to be perceptible.

To see why this is true, it is necessary to consider the actual
magnitude of Planck's quanta, of the steps by which the energy
of an oscillator may increase or decrease. This magnitude is
obtained by multiplying the *frequency* of the oscillator by a number known as *Planck's constant*, a quantity which will appear
again and again in the discussions of the tiny world of atoms. Its
numerical value, expressed in *erg-seconds*, a unit that corresponds
to energy expressed in ergs and frequency expressed in oscillations per second, is exceedingly small: 6.63×10^{-27} erg-second.
An erg is the amount of energy expended in raising a milligram,
about the mass of a single grain of sand, up one centimeter, truly
a minute exertion.

Using the customary symbol h for Planck's constant, the quantum of energy is $h\nu$, and the discrete values of an oscillator energy content E are:

$$E = n \times h\nu \qquad n = 0, 1, 2, 3 \dots$$

The quantum is thus *not* a definite amount of energy but is greater for oscillators of higher frequency. The frequencies of the tiny atomic oscillators are far higher than those of large-scale things like pendulums. The magnitude of the quantized energy steps is correspondingly greater and is therefore quite significant compared to the total energy involved. In the realm of visible, tangible objects the quantization is not perceptible; the smallest observable change of energy corresponds to the absorption or emission of huge numbers of quanta.

To give a crude analogy (with individual molecules likened to individual energy quanta): for a water tumbler so small that it could hold but ten molecules of water, its contents would obviously be "quantized," that is, restricted to certain particular values. It could be either empty or one-tenth filled or two-tenths filled, and so on, since no less than one whole molecule of water could be added or removed. This is true as well for a tumbler of ordinary size. But here the "quantization" is imperceptible.

With typical scientific conservatism Planck made no new assumptions about the nature of radiant energy itself. He believed, as did everyone at the time, that this energy is distributed smoothly and continuously through the space traversed by the waves. And he never faced the question of how his quantized oscillators could nevertheless interchange energy in discrete amounts with this continuum. Einstein, however, did deal with this problem when, in a paper published in 1905, he made the radical proposal that *radiant energy itself is quantized.* In terms of this concept a beam of light is a stream of tiny, speeding particles of energy. They are so minute and so numerous that they cannot be felt impinging individually any more than one can feel the impacts of individual air molecules in a gentle breeze.

For the amount of energy E in the individual particles Einstein used the same rule that Planck had applied to his quanta:

$$E = h\nu$$

These particles of radiant energy were later named *photons* in analogy with electrons, the particles of electricity. There is, however, the important difference that, while the quantity of electrical charge of all electrons is precisely alike, the energy content of photons is proportional to the frequency of the radiant energy.

The photon concept was used by Einstein to explain the *photoelectric effect,* the ejection of electrons from the surface of metals by incident radiant energy. On the basis of the classical wave theory it seemed that the intensity of the light, which determines the amplitude of the waves, should be the decisive factor in producing the effect. But experiments showed that the energy with which the ejected electrons fly off depends only on the *frequency* of the light, being in fact directly proportional to this frequency. This is easily explained on the assumption that each electron is ejected after having absorbed the energy of a single photon. For his conception of the photon and the successful theory of the photoelectric effect, Einstein was awarded the Nobel Prize.

This incisive new concept accounts readily for a great number of observations that had previously been quite inexplicable. Thus, red light has little effect on photographic plates while blue light is very active. Visible light is harmless to the skin but ultraviolet causes severe sunburn. Similarly, visible light, even when quite intense, does little harm to bacteria while X-rays kill them quickly. These observations and many similar ones are understood immediately in terms of the fact that red light, blue light, ultraviolet and X-rays form a sequence of diminishing wavelength and increasing frequency and, by Einstein's formula, of increasing photon energy. Since the wavelengths of X-rays are a thousandfold shorter than those of visible light, it is not surprising that their thousandfold more energetic photons are far more potent.

The one observation that demonstrates the photon theory most convincingly is the effect discovered in 1923 at the University of Chicago by Arthur Holly Compton (1892-1962) while investigating the scattering of X-rays. It had been known for several decades that this scattering is mediated by the atomic electrons. The classical theory assumed that, as seen in the Figure 9 the X-ray waves passing over the electrons e cause them to oscillate at the same frequency as that of the waves, and the oscillating electrons in turn emit X-rays in all directions. Through this two-step process some of the energy of the incident X-rays is scattered, and there is no possibility of any change in frequency or wavelength between the incident and scattered waves. Yet Compton made the curious observations that the scattered rays have longer wavelengths.

In terms of the photon theory this result is not only possible—it is inevitable. The reason is that, with photons, scattering is a kind of subatomic billiards. As shown in Figure 10 the incident photon of energy $h\nu_i$ collides with the electron e and ricochets off at an angle, the electron rebounding from the impact with considerable speed and energy. Since the photon thus contributes energy to the electron during the collisions, the scattered photon has less enegy, a lower frequency ν_s and longer wavelength. Compton was able to show by mathematical calculations that the observed shift in wavelength corresponds exactly with the predictions of the photon concept. The electrons in Compton's experiments were not actually standing free as shown in the diagram; rather, they were the outermost electrons of carbon atoms.

Figure 9. The scattering of X-rays according to the wave theory of light.

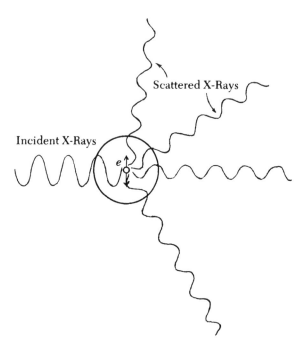

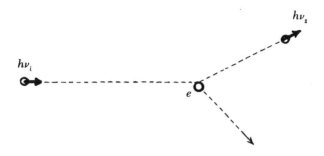

Figure 10. The scattering of X-rays according to the photon theory of light.

However, the small amount of energy required to detach these electrons from the atoms is negligible compared to the energy of the X-ray photons, and it has no appreciable effect on the results.

Since Young's interference experiment, together with a large body of other evidence, had led to the conclusion that light is a wave, while the phenomena considered here demonstrate that light behaves like a particle of energy, physics seemed to be left with a serious dilemma. The *dual nature of light* continued to exercise physicists for decades, and no generally acceptable solution was forthcoming. The British physicist Sir William Bragg (1862-1942) was moved to remark that scientists apparently have no other recourse but to believe in waves on Monday, Wednesday and Friday and in particles on Tuesday, Thursday and Saturday; and, he might have added, to pray for guidance on Sunday! As later discussions will show, the situation is not quite so intractable.

THE BOHR THEORY OF ATOMIC RADIATION

In the early decades of the twentieth century England had taken the lead in experimental physics, which was of primary interest to its leaders in research, but theoretical studies had not been emphasized. On the Continent, however, and particularly in Germany, theory had made important advances, in good part through the work of Planck and Einstein.

It was an auspicious circumstance that young Niels Bohr came under the influence of both developments. When he arrived at Rutherford's laboratory, the new ideas of photons and quantization of energy were fresh and vivid in his mind; and he addressed himself to the problem of relating these concepts to the radiation of light by Rutherford's planetary atoms. Planck had already studied the interaction of radiation and material substance. But whereas he had dealt with the atoms *en masse*, Bohr pinpointed the problem by considering the interaction with single atoms, in particular with the simple atoms of hydrogen.

When hydrogen is excited to radiation in the electric discharge tube, it emits light consisting of a complex array or *spectrum* of certain particular colors or wavelengths. Some of these lie in the visible range, others in the infrared and ultraviolet, the whole spectrum containing well over a hundred distinct wavelengths.

The classical theory of radiation had associated the emission of any one wavelength or frequency of radiant energy with the existence within the atom of an elastically bound electron, an electronic oscillator, that is set in motion by violent agitation of the atoms, such as occurs in the electrical discharge. The difficulty in attempting to account for the hydrogen emission spectrum lay in explaining how this simple atom with its single electron could contain over a hundred such oscillators, each one "tuned" to one of the frequencies which the atom is seen to emit. Arnold Sommerfeld (1868-1951), the director of the Institut für theoretische Physik in Munich, who was not only a great scientist and teacher but also an amateur pianist of no mean ability, remarked wryly that the hydrogen atom appears to be more complicated than a grand piano (which emits only eighty-eight distinct frequencies).

In attacking this problem, Bohr dismissed the idea of atomic oscillators; nor did he attempt, as others had done unsuccessfully before, to relate the frequencies of the emitted radiation directly to corresponding motions of atomic electrons. Rather, he proposed a completely different mechanism for the emission of radiant energy. With particular regard to hydrogen atoms, his basic postulate is that the lone electron may circulate about the nucleus, not only in its normal orbit, in shell number 1, but may *jump* to any one of a sequence of *larger* orbits. Any transition to a wider orbit requires that a supply of energy enter the atom to move the electron outward against the attracting force of the nucleus; conversely, a jump to a smaller orbit, nearer to the nucleus, must be accompanied by ejection of energy from the atom. This, in brief, is Bohr's conception of the absorption and emission of radiant energy by atoms.

Bohr pictured an atom in the electrical discharge tube as suffering a collision sufficiently violent to hurl its electron from the normal to a far-flung orbit, from which it returns, in one or in several jumps, to the normal state, emitting energy at each jump. He assumed that the emitted energy is in the form of Einstein's photons. The problem was to account for the quanta of energy in these photons, corresponding to the various particular frequencies emitted by the hydrogen atom. According to classical laws, the electron can circulate about the nucleus in an orbit of any size whatsoever just so its velocity has the proper value for that orbit. A *continuous* range of orbit sizes would not do, however, to explain the observed emission spectrum; it is necessary to have a sequence of *quantized* orbits, each with a discrete amount of energy. Downward jumps between them thus result in the emission of photons with discrete amounts of energy, corresponding to the *difference* in energy of the two orbits between which the jump occurs, that is, the loss of energy by the atom is equal to the energy carried off by the emitted photon. Bohr's problem was to find a general *quantization condition* that would assign the proper energy values to all the orbits.

After giving this matter much thought, he arrived at the following relation:

$$m\,v\,r = n \times h/2\pi \qquad n = 1, 2, 3 \ldots$$

Here m, v and r designate, respectively, the mass and the velocity of the electron and the radius of the orbit in which it travels, h is Planck's constant, and π (pi) has its usual significance as the ratio of the circumference to the diameter of a circle. Of greatest significance is the appearance of the natural numbers 1, 2, 3 . . . , this being the second instance in which they occur in the formulas of atomic physics. Their meaning here is that the product of the three factors at the left of the formula (a product known as the *angular momentum* of the electron) is just equal to $h/2\pi$ or just two or just three times this value, and so forth. This sequence of discrete values of the angular momentum results in a corresponding sequence of discrete values of the orbital radii, no sizes in between being possible.

As a first minor triumph, the quantization condition, with n set equal to unity (for the smallest orbit), gave a value for r in close agreement with the known size of the hydrogen atom in the normal state (10^{-8} centimeter). But of far greater moment, when Bohr calculated the sequence of energy values corresponding to the orbit sizes given by the quantization condition, he obtained precisely the values needed to account for the observed emission spectrum. This was a momentous achievement that gained universal acclaim.

Bohr delineated this whole process in the form of an *energy-level diagram*, as shown in Figure 11. Here E_1 is the ground-level energy, and the rest of the levels correspond to the higher quantized energy values of successively larger orbits. The step from E_1 to E_2 is the greatest, the rest becoming progressively smaller. The levels toward the top are too crowded to be shown in the diagram. But careful experiments revealed a number of energy levels sufficient to provide easily for more than a hundred transitions between them. The big jumps down to the lowest level correspond to the emission of photons having high energy, which

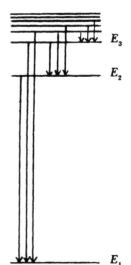

E_3

E_2

Figure 11. Bohr's energy-level diagram for the hydrogen atom.

E_1

means high frequency and short wavelength. This radiation is therefore in the ultraviolet region of the spectrum. Jumps down to level E_2 correspond to photons of visible light, while the low-energy photons emitted during jumps to E_3 constitute infrared radiation.

By this simple mechanism Bohr accounted for the great complexity of the hydrogen spectrum. He expressed the relation between orbital energies and emitted frequencies by the formula:

$$h\nu_{2,1} = E_2 - E_1$$

Here $\nu_{2,1}$ is the frequency of the radiation emitted during the jump from level E_2 to E_1, and $h\nu_{2,1}$ is the energy of the photon. The same relation may be written for all the other transitions.

With proper experimental conditions it is possible to observe the *absorption* of photons by the atoms. Here the transitions go from lower to higher energy levels, but this same relation holds.

Once Bohr had led the way, many others joined him in elaborating and extending his theory to describe the spectra of the other, more complicated atoms. Their electron orbits were assumed to be quantized; and their spectra were explained by assuming that one of the electrons jumps to a more far-flung orbit when energy is absorbed and drops back to a smaller one when energy is emitted, as in the hydrogen atom.

Each kind of atom is found to have its unique array of energy levels. This is in agreement with the observation that each emits its own specific spectrum of wavelengths. The optical spectrum thus provides a unique means of identification; and it establishes with complete certainty the presence of a particular kind of atom in any source of light, be it here on earth or in a distant star.

The Bohr atom model was thus spectacularly successful in accounting both for the spectroscopic data pertaining to the interaction of atoms and photons and for the observed chemical behavior of atoms, for their interaction with one another. These two areas of experimental observations could be related when it was realized that the successive electron shells correspond to the sequence of quantized orbits of the hydrogen atom, though the sequence of energy values associated with them is greatly altered by the presence of the additional electrons. Even the stability of atoms against collapse of the electrons into the nucleus was explained formally by the concept of quantized orbits, with one of minimum size for each electron.

Yet there were disquieting difficulties. As spectroscopic measurements were made more precise, certain fine details were found to be definitely at variance with the Bohr theory. In a study made in 1925 to explain these discrepancies, the two young Dutch physicists Samuel Goudsmit and George Uhlenbeck came to the conclusion that atomic electrons are *spinning* about their axes while traveling around the nucleus, much as the earth turns about the poles while making its tour around the sun. Although it was only partially successful in explaining the puzzling spectroscopic details, this idea was found later to have far-reaching consequences.

Of equal concern was the inability of the Bohr theory to account for the *intensity* or brightness of the various colors of light emitted by atoms. Intensity is certainly an observable attribute of light, and an adequate theory should be able to explain it. In the Bohr atom model the intensity of light emission is to be correlated with the nature of electron jumps to orbits of lower energy; jumps that occur more readily and frequently should give rise to more copious emission. But the Bohr theory says nothing about transitions. The electron simply is first in one orbit and later in another without having passed between them in any describable manner.

Nevertheless, some physicists, setting aside these difficulties, were led to surmise that the Bohr atom model, because of its many positive achievements, might well be a close approximation to the "real" atom. But Bohr himself, in his publications and speeches and private conversations, continued to emphasize the tentative character of his concept and, in particular, its logical inconsistencies. In his calculations he had applied the classical laws of mechanics and electrodynamics where they were useful, but had violated these same laws in his postulates of quantized orbits and abrupt transitions between them. In common with other thoughtful physicists he hoped that the deeper truths, which might explain the remarkable measure of accord with the facts of experiment, would eventually be laid bare.

This hope was to be realized through ways which shall be considered in due course. But first it is requisite to clarify the light-wave-versus-photon dichotomy and to note an important consequence of it.

5

Consequences of the
Wave-Photon Concept

In the discussion of radiant energy just concluded, it was shown that both the streaming of light along straight lines when passing large objects, and the departure from a rectilinear course when it encounters objects of very tiny size may be explained in terms of its wavelike nature. It is a property of all sorts of waves, including sound waves, ripples and ocean waves, to proceed straight ahead past objects that are large relative to the wavelengths but to wash around objects whose size is comparable to this length. In particular, the optical experiments of Thomas Young gave convincing evidence for the wave nature of light.

On the other hand, it appeared that there are numerous phenomena which may be explained readily on the assumption that light is a stream of tiny energy packets, the photons, the energy content of each packet being inversely proportional to the wavelength (or directly proportional to the frequency). This concept accounts for the greater potency of short-wave radiant energy, ultraviolet light and X-rays, as compared to visible light and the infrared. The X-ray scattering experiments of Compton and Bohr's account of atomic spectra give strong support for this particlelike conception of light.

As noted before, the whole of this experimental evidence leads to an apparently irreconcilable dichotomy: light appears to be both a wave and a particle. This difficulty may be resolved by examining carefully the nature of the experimental evidence in support of both views. Here it is expedient to consider once again Young's experiment on interference of light waves. Using modern equipment not available in his time, it is possible to carry out the experiment in such a manner as to exhibit both the wavelike and the particlelike properties of light and to arrive at conclusions concerning their relation.

The essential new device is a detector of radiation so sensitive that it can perceive the incidence of a *single* photon. In this *photomultiplier* the photoelectric ejection of a single electron sets off a spurt of about a million additional electrons. This constitutes a pulse of electric current which may be enhanced further by a conventional electronic amplifier to produce an audible "click" when it passes through a radio speaker.

As this detector is drawn slowly across the pattern of bright and dark bands on the screen of Young's apparatus (page 47), the response heard in the speaker alternately increases and decreases. With the detector at a dark band, only an occasional click is heard. As it moves into a bright band, the clicks rise to a crescendo of hundreds per second and subside again at the following dark band. Thus, this single experiment reveals both the arrival of individual photons at the screen and the pattern of bands which results from the action of waves.

These observations, and many others of a like nature, have been explained by considering that all the *energy* of the light is carried by the photons, and that their flight is *guided* by the waves in that they flock in large numbers to those places where the wave concept implies bright light and avoid the places where the light is dim. More precisely, the number of photons at any given place is proportional to the *intensity* of the light determined for that place on the basis of the wave theory.

Quite generally, experimental studies concerning the nature of light may be directed at either one of two aspects. They may pertain to the *propagation* of light, to its passage through space or through various optical systems; or they may be concerned with *emission* and *absorption* processes, with the way in which light arises out of the energy of material substance and changes back again from the radiant form. Observations of the absorption and emission of photons by atoms is an important example of the latter. Studies of how a rainbow is formed in the atmosphere or how a camera lens forms an image are of the first kind. Experiments in both of these areas show that all propagation phenomena may be explained by the wave theory while absorption and emission are accounted for in terms of photons. This general observation is in accord with the concept of photons guided by waves.

The very act of *seeing* involves both aspects. The wave concept is needed to trace the beam of light from the observed object through the lens of the eye and onto the retina. Here the light is absorbed, as a packet of energy, and initiates a complex sequence of physiological and psychological events resulting in the sensation of sight.

The relation of photons and waves may be indicated by an analogy in mechanics. When a stone is tossed over a field, it follows a certain curved path called a *parabola*. This parabolic path is certainly "real" in the sense that it is the actual path taken by the stone, and it is also "real" insofar as it is contained in the *equations of motion* calculated mathematically by applying the laws of Newtonian mechanics to the flight of the stone. But it need not be thought of as being a real *physical* entity.

Similarly, light waves may be looked upon as a purely mathematical concept, a pattern in space given by the laws of wave optics, which determines the path of the photons but has no real physical existence. This analogy is to be accepted with caution, for, as will become apparent later, there is a profound difference between the way in which the laws of classical mechanics determine the path of the stone and the manner in which the photons are guided by the laws of wave propagation.

With this interpretation of light waves it is possible to resolve some apparent difficulties regarding the behavior of photons. If, for example, one of the two slits in Young's optical apparatus is closed, light streaming through the open slit is distributed rather smoothly over the whole expanse of the screen. With both slits open, there is the previously described distribution in light and dark bands. Photons going through the open slit distribute themselves in two quite different ways, depending on whether the other slit is closed or open. The question arises as to how photons going through one slit "know" whether or not the other is open. The answer is that the wave patterns calculated for one and for two open slits are quite different, and the photons follow the calculated pattern in either case.

It has been argued that light waves must be accorded a measure of reality because the very expression for the photons' energy ($h\nu$) involves the *frequency*, a concept that pertains to waves and has no meaning for particles. This argument can be shown to be of little value by the sound procedure of separating fact and fancy, by distinguishing carefully between what may actually be observed and what is added as a mental conception. Experimentally, it is possible to determine the *energy* (E), which a photon gives up as it is absorbed (for example, by measuring the energy of an ejected photoelectron) and the *place* at which this energy exchange takes place. From the measured value of E it is possible (using the relations $E = h\nu$ and $\lambda = c/\nu$) to calculate a numerical value for a quantity λ which need not be accorded any physical significance beyond this mathematical relation to E. By introducing this numerical value of λ into a mathematical analysis appropriate for the optical situation at hand, a spatial pattern is obtained which predicts correctly the distribution of the energy packets. Expressed more briefly, the distribution pattern of the photons in any given optical system is determined by their energy. In this mathematical analysis it is not at all necessary to use the words "frequency" or wavelength."

It might be conceded that waves are a convenient mental construct for visualizing the distribution of the photons (as was done in the discussion of Young's optical experiment). There is no experimental justification for attributing any energy to the waves; and, bereft of all energy, they cannot possibly be the direct cause of any observable effects. Einstein, who had reduced the waves to this state by his conception of photons, dubbed them "ghost waves."

Many physicists did not take kindly to the idea of relegating light waves to the role of mere "ghosts," for during all of the nineteenth century, following the work of Young and Freznel, a succession of eminent scientists had produced a wealth of experimental evidence in favor of the wave concept. Most of these experiments dealt with the *propagation* of light, in which its wavelike properties would be brought to the fore; and such studies as were made of absorption and emission phenomena were concerned only with gross average effects, which could also be described in terms of waves. Thus, as the nineteenth century drew to a close, nothing in the whole realm of physics seemed more thoroughly substantiated than the conception that light is a wave. Although in the early decades of the twentieth century the existence of photons was verified by convincing experimental evidence, the strong sense of reality, with which long familiarity had invested the waves, was not easily dispelled. The conflict of the two irreconcilable ideas led to the apparent dilemma of the "dual" nature of light.

It would appear that this problem of duality is in part a semantic one. When using the word "wave," it is difficult to avoid intuitive associations with familiar phenomena, such as waves on water, whose energy is carried along in their undulations and is spread out smoothly over the region through which they travel. Similarly, the word "particle" suggests things, like grains of sand, which move in accord with the laws of classical mechanics. If light were found to possess all the properties of such familiar "waves" and "particles," its nature would indeed be "dual"; it would be contradictory and incomprehensible.

Actually, light is a distinct entity whose properties, in part wavelike and in part particlelike, have been studied thoroughly and are well understood. Furthermore, each of its two aspects is evident in one of two distinct categories of observation; therefore, they never clash experimentally. And when rightly understood, they need not clash logically either. Thus, it is quite within reason to think of both waves and particles on every day of the week.

Photons are particlelike in one other important respect: it is quite possible to think of them as moving through space that is quite empty, devoid of any transporting medium. Divesting light waves of physical reality disposes of the difficulties engendered by the concept of a luminiferous ether, for if nothing is actually undulating, there is no need of a medium to carry the undulations.

Finally, it is appropriate to note that the photon concept is by no means a return to the eighteenth-century idea of "light corpuscles," which had prevailed before the wave theory was accepted. These corpuscles were thought to behave according to the laws of Newtonian mechanics while photons clearly behave otherwise.

CAUSALITY AND PROBABILITY

THE concept of photons guided by waves provides a satisfactory account for the experimentally observed behavior of light, both with regard to its propagation and its interchange of energy with material substance. But this resolution of the "dual" nature of light has a most remarkable consequence. Consider, for example, a single photon that has traveled from the light source in Young's optical interference apparatus to one of the two slits. As a moment's consideration shows, it is not possible to predict how it will go on from there. All that can be said is that the photon follows the course of the waves, and it can proceed to various points on the screen beyond the slits. Furthermore, if the photon is detected at some point on the screen, there is no way of knowing from which of the two slits it has come. Thus, the concept of a *path* followed by the photon from slit to screen becomes quite vague.

This is a situation completely at odds with the spirit of classical science. Since the days of Descartes and Newton scientists had regarded the physical universe as a system in which all events proceed in an unbroken chain of cause and effect, according to a universal *principle of causality*.

In a universe in which the principle of causality holds, the laws of nature are *deterministic*, the experimental verification of determinism being provided by the *predictability* of events. The principle of causality asserts that, if all the pertinent facts concerning the present state of a system are known, it is possible by means of the laws of nature to predict its state at any later time.

When a gunner fires a cannon to hurl a projectile with a certain speed and in a certain direction, he expects it to follow a definite path to the target according to the predictions of the laws of mechanics. If the path is found to be not precisely as predicted, he lays this to inadequate control of the conditions under which the projectile was fired; and he is convinced that he can make the path conform more closely to the predicted one by controlling these conditions more carefully.

All of experimental science and technology, and much of the conduct of everyday affairs, presupposes the predictability of events by deterministic laws. It is predictable that a spoken message will be transmitted intelligibly over a properly designed telephone line and that an airplane designed on well-established aerodynamic principles will fly, under known conditions, properly and at a predicted speed. A kettle of cold water placed on a hot stove will surely get hotter as the laws of heat flow predict; and an automobile will slow down when the brakes are applied, in accord with the law of mechanics.

Prediction is, however, quite impossible for the photon in Young's apparatus. Its motion after passing through the slit is not determined by its condition immediately before it reached the slit. Two photons could pass from source to slit in the same way and yet take quite different paths thereafter. In particular, no effort at making their passage to the slit more nearly alike could improve the prediction of their subsequent behavior.

This remarkable observation is quite general for optical phenomena. It is impossible to predict the behavior of individual photons no matter how carefully conditions are controlled. What is involved here is apparently nothing less than the *breakdown of the principle of causality*. (This conclusion will be examined more critically in chapters 16 and 18.)

This new insight should have been profoundly disturbing. But actually, the majority of physicists were not greatly concerned, for the breakdown did not occur so long as light was thought of as a phenomenon of waves; light waves behave quite in accord with the deterministic laws of wave optics. It was the photons which failed to conform, and photons were new and strange and controversial while light waves were familiar and well established. Furthermore, for many practical purposes it is not necessary to determine the paths of individual photons; the knowledge of their overall group behavior is sufficient.

This group behavior may be related mathematically to the distribution of the waves by means of the concepts of *probability* and the laws of *statistics*. Statistical laws were first thought of in relation to games of chance. When a coin is tossed, it has an equal chance of showing heads or tails. Therefore, the laws of statistics predict that in 2000 throws the count should be very nearly 1000 heads and 1000 tails. Statistical laws can make no predictions about the outcome of a single event nor are they reliable for a small number of them. In 10 throws heads might well show 7 times; but it is highly improbable that this would happen again in the next 10, or the next and the next. Therefore the probability of finding 700 heads in 1000 throws is practically nil.

In Young's optical experiment (page 46) the waves may be thought of as a spatial *pattern of probability*. If the calculated intensity of the waves at a point y on the screen is twice as great as at another point x, the probability of photons arriving at y is also twice as great as at x. Over a sufficiently long time, during which very many photons fall upon the screen, the number actually arriving at y will be very nearly double that at x. Thus, while the wave pattern affords no prediction concerning the behavior of individual photons, it does predict their statistical distribution over the screen with a high degree of precision. This is generally true in the many example of optical phenomena that involve great numbers of photons.

Those familiar with the developments of nineteenth-century physics will note that there is nothing new in dealing with overall average effect of large numbers of individual events by statistical methods. Such methods were used for describing the observable properties of gases in terms of the statistical behavior of their component molecules. There is, however, a basic difference between the way in which statistical factors enter into the behavior of gas molecules and of photons.

To make this distinction evident, it is useful to consider an easily visualized system: a number of balls moving about on a billiard table. To simplify matters, it will be assumed that all collisions of the balls, with one another and with the edges of the table, are perfectly elastic so that once set in motion they continue to move with undiminished aggregate energy. If there is but one ball, it is easily followed; and from the observation of its position and motion at one given instant its future course may be predicted with certainty by means of the laws of mechanics. If there are two balls, prediction becomes considerably more difficult since the motion of both is affected by mutual collisions. As the number of balls grow larger, the difficulty increases greatly; with a hundred, prediction is practically impossible. Yet for still greater numbers prediction of their statistical behavior becomes progressively more reliable. With 2000 balls milling about on the table, it is highly probable that very nearly 1000 will be at any moment on either half of its surface.

When gases were first studied experimentally as "tenuous continua," the laws governing their behavior were thought of as being truly deterministic, much like the laws of classical mechanics. The earliest of these laws, discovered in 1662 by Robert Boyle, asserts that, when a gas is compressed to half its original volume (while maintaining its temperature constant), the pressure is doubled, at one-third original volume it is trebled, and so forth. With the development of the kinetic theory of gases (page 17), this observation was explained in terms of the statistical behavior of the gas molecules: when crowded into half the original volume, they collide with the walls of the containing vessel twice as frequently, thus producing twice as great a pressure. Because of the enormous number of molecules involved, their statistical behavior becomes highly predictable. Therefore, an engineer dealing with the behavior of gases, as in the design of an air compressor, need not concern himself with the random behavior of single molecules; he can depend on the apparently deterministic statistical behavior.

Similarly, the designer of a camera or a telescope need not consider the behavior of individual photons, for here again the overall statistical behavior, as described by the laws of wave propagation, is highly predictable. Yet, in spite of this apparent similarity, there is a fundamental difference between the two situations. In the case of gases statistical methods are employed because the large numbers of molecules involved make it impossible to carry out a detailed analysis. For the photons statistical considerations enter in a more fundamental manner. Although it is entirely possible to make the light intensity in an optical experiment so low that there is only one photon present at a time in the entire apparatus, it is nevertheless impossible to determine the behavior of this single photon. For the very laws to which its motions are ascribed can give no more than statistical predictions.

After this long digression into optics it is now in order to resume the discussion of mechanics and to consider some radically new developments for which the ideas just presented will be found relevant in unexpected ways.

6

Matter Waves

A NEW MECHANICS

CONCEPTS, principles, theories and laws—these are the tools of the theoretical physicist, tools which he must fashion as he goes about his job of describing the workings of nature. The tools of classical physics had been developed in dealing with the familiar world of large objects. When the work of Thomson, Planck, Einstein, Rutherford and others carried the center of interest in physics over into the submicroscopic realm of the atoms, it became increasingly apparent that the old tools were not suitable for the new job. As the discussion of Bohr's atom model has shown, new concepts had to be introduced to account for the experimental observations. This in itself was not too disturbing; it was rather to be expected that the old laws would need revising. But difficult to accept were the logical contradictions between these new ideas and those of classical physics. These difficulties were, however, inevitable, for the essential new property which came to the fore in the atomic realm was *discontinuity*, abrupt change of energy and of form, and such discontinuity is foreign to the basic concepts of classical mechanics.

Although this innate source of the problem was not generally discerned, there arose a growing conviction that no amount of patching up of old laws would do, that a fundamentally new approach was needed to set things aright. What form this new

approach might take or when it would come about was impossible
to foretell. But there was a hopeful consensus that, as on previous
occasions when some area of science had faced an impasse, someone would come forth with a conception which would lead to
renewed progress.

The anticipated new approach came about when, in 1924,
Professor Paul Langevin (1872-1946) of the Ecole Polytechnique
in Paris received from one of his students a rather puzzling
doctoral dissertation containing elements of the mathematical
methods of classical mechanics and optics, together with ideas
taken from Einstein's theory of relativity and the quantum theories
of Planck and Bohr. The author of this dissertation, Prince Louis
Victor de Broglie, had come to the study of physics in an uncon-
ventional manner. Born of a noble family with a long tradition of
scholarly achievement, he began his studies at the University of
Paris in ancient history and paleography. But through the influence of his older brother Maurice, who was doing notable work
in the field of X-rays and radioactivity, he acquired an interest in
theoretical physics. He applied himself to the urgent fundamental
problems of atomic theory and arrived by a brilliant feat of intuition at a concept that led to the development of a new and
logically consistent theory of mechanics.

To understand how de Broglie came upon his novel idea, it is
appropriate to consider once again the optical apparatus of
Thomas Young, using, as before, light of one particular wavelength
but making the two parallel slits adjustable in width. With wide
slits, for which diffraction effects are negligible, the light from
the source streams straight through, as shown at (1) in Figure 12,
and illuminates two well-separated areas B and B'. As the slits are
made narrower (2), as their width approaches the wavelength of
light, diffraction causes the bright areas to spread and overlap
(3); and the interference of the light from both slits produces a
pattern of light and dark bands.

An eighteenth-century physicist, thinking of light as a stream
of corpuscles moving according to the classical laws of mechanics,
would find the wide-slit pattern of illumination quite comprehensible. He would be astonished, however, at the behavior of the
corpuscles passing through the narrow slits; and he would conclude that for some unknown reason the laws of mechanics had

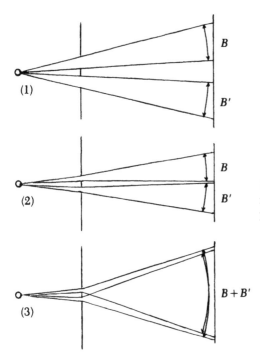

Figure 12. Paths of light rays through Young's apparatus.

ceased to hold. The same inexplicable behavior would be observed whenever the corpuscles stream past any very tiny objects. For this there is an obvious analogy in the mechanical behavior of material objects; they too obey the classical laws in large systems but fail to do so in very tiny ones.

All observations on the propagation of light had been fully explained during the nineteenth century in terms of the wave concept; and when, in the twentieth, light was again endowed with particlelike properties, waves were still retained in the wave-particle "duality." Pondering these circumstances, de Broglie was led to speculate whether, conversely, material particles might not have some wavelike properties, whether like photons they might not be *guided by waves*. Though no one had ever imagined that waves might be involved fundamentally in mechanical phenomena, he decided nevertheless to explore the consequences of this strange notion.

De Broglie could establish a relation between photons and material particles using Einstein's well-known formula $E = mc^2$, which states that an amount of energy E is equivalent to an amount of mass E/c^2. Since the energy of a photon is $h\nu$, its equivalent mass m is $h\nu/c^2$. Using the relation $\nu = c/\lambda$, this may be written $m = h/\lambda c$ or $\lambda = h/mc$, that is, for light the wavelength equals Planck's constant divided by the product of mass and velocity.

De Broglie suggested that an analogous relation might hold for the "matter wave" associated with a moving particle, that here also the wavelength might be equal to Planck's constant divided by the product of the particle's mass m and its velocity v:

$$\lambda = h/mv$$

This is the celebrated *de Broglie matter wave relation*.

The product mv, which is of great importance in many mechanical phenomena, is called the *momentum*, usually designated p. (It was formerly called "quantity of motion," an apt name, in keeping with the intuitive feeling that there is a great deal of motion to a body that is massive and is moving rapidly.) The de Broglie wavelength of a particle is thus in inverse ratio to its momentum.

Wavelength, as previous discussions have shown, is of primary significance in determining the behavior of any kind of wave. Therefore, it is important to look into the numerical value of λ given by the de Broglie formula. For an object having a mass of one gram and a velocity of one centimeter per second (typical values for an object of familiar size), λ has the numerical value of h, about 10^{-26} centimeter, an absurdly small value, ten million million times smaller than an atomic nucleus. Thus, in the familiar large-scale world all objects are of enormous size compared to their matter waves; this explains immediately why diffraction effects are absent and no departure from classical laws is found. Those familiar with mathematical relations will note that in the de Broglie relation λ becomes larger as v is made smaller and that, by allowing the one-gram mass to move very slowly, λ could be increased to a value at which departure from classical behavior would become evident. However, this would require a reduction of v to an extremely low value, perhaps a hair's breadth of travel in billions of centuries, hardly an observable "velocity."

At the level of atoms the situation is quite different. From Bohr's well-substantiated quantization condition (page 57) it follows that, for the electron of the hydrogen atom in its normal state, the momentum mv is $h/2\pi r$, giving for λ the value $2\pi r$, the circumference of the electron orbit. Roughly the same relation holds for all atomic electrons, that is, the matter waves of the electrons are about equal to the size of the system in which they are moving. This is just the condition under which strong diffraction effects, and marked departure from the classical laws of mechanics, should be expected.

While the results of these simple numerical calculations lend a measure of support, they could not be construed as a proof of de Broglie's remarkable idea. However, direct experimental evidence for the existence of matter waves was forthcoming in little more than a year's time after they had first been proposed. In fact, such evidence was being provided, albeit unrecognized, even while de Broglie was engaged in his speculations.

Two young American physicists, Clinton Davisson (1881-1958) and Lester Germer, working in the laboratories of the Bell Telephone Company, had been studying the behavior of streams of electrons impinging on the smooth surface of a crystal of nickel. They expected that the electrons would rebound from the surface like elastic spheres and that they would do so equally well at all angles of incidence. Instead, they found that the rebound was strong at certain particular angles and relatively weak at all others. Further, the angles of strong rebound were observed to change in a definitive manner with a change in the velocity of the impinging electrons. These observations were inexplicable until Davisson, while attending, in 1928, a meeting of physicists in London, heard about the de Broglie matter waves.

The reflection of waves from crystal was well understood, for it had been investigated as far back as 1912 through experiments with X-rays. The atoms in a crystalline material are arranged in rows and columns to form flat layers, one upon the other. Just two such layers are indicated by the heavy dashed lines in Figure 13. The laws of wave propagation show that X-rays are reflected from a flat layer of atoms as visible light is reflected from a plane mirror. X-rays penetrate into solid materials and are therefore reflected not only at the surface but on a number of parallel layers underneath. Thus, the ray r is reflected on the

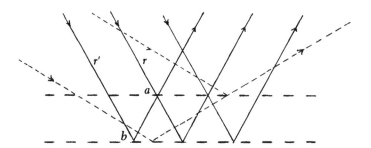

Figure 13. Interference of X-rays reflected from two successive layers of atoms in a crystal.

surface at *a* while *r′* is reflected at *b*, on the next layer down, and joins *r* at *a*. If the angle of incidence of these rays is such that the additional distance traveled by *r′*, while passing down to the second layer and back, is equal to just one wavelength, the two rays meet at *a* in phase and add to form a strong wave. This same reinforcement occurs for all the rays reflected at the successively deeper layers; therefore, at this particular angle of incidence there will be strong total reflection. For rays coming in at other special angles, as indicated by the fine dotted lines, for which the additional distance of travel between layers is just two or three or any whole number of wavelengths, there will be the same reinforcement and strong reflection. But at all other angles the phase relations between the rays coming from successive layers will not be right for reinforcement and reflection will be weak. For X-rays of various wavelengths the angles of strong reflection are, of course, different; and if the spacing of the atom layers is known, the wavelength can be computed from the observed values of these angles.

Obviously, if electrons are guided by matter waves, they too should rebound copiously only at the particular angles at which these waves are strongly reflected. When calculations were made of the wavelengths corresponding to these measured angles, they were found to be in agreement with the de Broglie formula. Further, the different angles found for electrons of various velocities were in accord with the relation between λ and v given in this formula.

A year later the English physicist Sir George P. Thomson, son of the discoverer of the electrons, shot beams of electrons through thin plates of crystalline materials and found that here also they were scattered only in certain angles determined by the wavelengths of their matter waves. Others studied the reflection on crystals of protons, of neutrons and even of atoms, thousands of times more massive than electrons. Again, observed motions were found to be in agreement with the matter wave concept and not at all as predicted by Newton's laws, for these could not possibly account for motions in certain particular directions only.

All attempts to explain these observations by less radical conceptions proved in vain. Physicists had to accept the astonishing idea that waves are involved in the motion of these tiny particles.

For objects of greater mass, as the de Broglie wavelength relation shows, the wavelength is correspondingly shorter. A smaller change in the angle of the rays suffices to increase the additional distance of travel between layers by one wavelength. This means that for more massive particles the successive angles of strong reflection are more closely spaced. If a series of experiments on crystal reflection could be performed, first with electrons and successively with particles of greater mass, the special angles giving strong reflection, at first few in number and widely separated, would become successively more numerous and more closely spaced. Finally, using spheres of visible size many billions of times more massive than electrons, these special angles would be far too close to be separately distinguishable. There would be perceptibly a continuous range of angles, all of equally strong reflection; and the particles would be found to rebound equally well at all angles, apparently in accord with the laws of classical mechanics. Nowhere in this series of experiments would there be a sudden break; the transition from the discontinuous group of discrete angles observed for entities at the atomic level to the continuous behavior observed with things of substantial size would be quite smooth.

What is seen in this particular situation is true in general for the mechanical behavior of objects. Both the large and the very minute move in accord with the new laws of wave propagation. But for large things the implications of these new laws become identical, to all semblance, with those of the classical laws. Nevertheless, it cannot be disputed that the laws formulated by Newton, which had for two hundred years been thought universally valid, are seen to hold only for things of large size, only for a part of the whole range of sizes encompassed by the more general laws based on de Broglie's concept of matter waves.

Why this conclusion was so long in being recognized is again a matter of wavelength. Light waves are indeed minute; yet there are things in the familiar environment, such as tiny pinholes and fog droplets, that are almost as small. Thus, light waves were discovered as soon as the interaction of light with such tiny objects was given careful attention. But matter waves are, in general, so inconceivably small that only for the very lightest of particles interacting with the atomic crystalline structure did their existence become evident.

There remains the question of the nature of these strange new waves. While physicists have by no means been of one mind in this matter, it seems possible to think of them, in analogy to the "ghost" waves of light, as devoid of real physical existence, as nothing more than a pattern in space and time determined by the mathematical laws of wave behavior. Here this concept is more easily accepted than it had been in the case of light, for these new waves are not burdened with preconceptions of physical reality.

The observation made for light waves that they hold to a straight course (through one specific kind of transparent medium) when diffraction effects are absent is true for matter waves *only* in situations in which the moving objects associated with the waves do themselves travel straight ahead, that is, when there are no *forces* acting to deflect them. Ordinarily, moving objects are not free of forces; a stone tossed up at an angle takes a *curved* course back to earth because of the downward pull of gravity. In terms of de Broglie's concept this occurs because deflecting forces cause matter waves to travel in curved paths. These curved wave paths are easily explained by the de Broglie relation between wavelength and momentum. The effect is analogous to the *refraction* of light, the bending of light rays in passing between two different transparent substances, as between air and water. (This is quite unrelated to *diffraction*, the bending of light rays around tiny objects.) Refraction is easily demonstrated by thrusting a pencil at any angle through the surface of water in a full drinking glass. The pencil appears bent at the surface of the water because light rays coming up through the water are deflected as they pass into the air. Refraction is caused by the fact that light travels at different velocities in various media, fastest in empty space, slightly slower in air, and only about three-fourths as fast in water.

The behavior of matter waves under the influence of deflecting forces may be likened to the refraction of light in the atmosphere when it passes between layers of air that are at different temperatures. The fact that light waves travel faster in warm air than in cold causes them to take a curved path, as shown in Figure 14.

The figure at (1) illustrates the formation of mirages in the desert. During the day there is a layer of warm air directly over the sand, with cooler air above. The row of thin lines indicates the successive positions of the front of a light wave coming down from the sky at a glancing angle. This wave front progresses more rapidly near the ground and follows the path indicated by the heavy curved line. Since an observer judges the location of what he sees to be in the direction from which the light enters his eye, he believes that the patch of blue sky is located at the point on the sand indicated by the dotted line; and he is apt to mistake it for the blue waters of a lake.

Mirages may appear in the sky under meteorological condi-

Figure 14. Production of mirages.

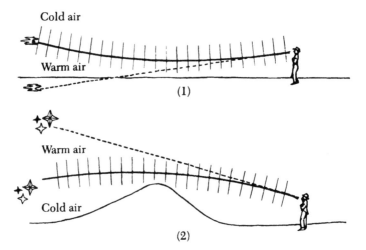

Cold air

Warm air

(1)

Warm air

Cold air

(2)

tions where there is a layer of warmer air above a cooler one near the ground. At night, as shown at (2), streetlights hidden from direct view by a hill may seem to be in the sky. Since the air layers are usually in motion, the mirage may also appear to move. This is presumably an explanation for some observations of mysterious flying objects.

For a tossed stone the velocity, and the momentum, decreases as it rises to the crest of its path and increases again on the way down. By the de Broglie formula (page 73) smaller momentum implies greater wavelength. The length of the waves, and consequently their velocity, is greater at higher points of its path and less at lower points, analogous to the conditions producing the mirage in the sky; and the wave path is curved in a similar manner. The curved course of the waves determined by these considerations is identical with the curved path of the stone given by the laws of classical mechanics.

This is generally true for bodies of substantial size. Wherever forces are acting, the guiding matter waves are affected in such a manner that the changes in velocity, whether they be a speeding up, a slowing down or a turning away from a straight path, are sensibly the same as those given by classical laws. But the wave concept is capable of describing as well the motion of very tiny things in very tiny structures, as that of electrons impinging upon a crystalline arrangement of atoms.

PARTICLES AND WAVES

THE conclusion that material particles are guided, like photons, by waves leads immediately to the consequences that the motions of individual objects *are not subject to deterministic laws*.

This qualification of the principle of causality had already been encountered with photons (page 67), but scientists took it more seriously when it became evident for electrons.

With light, the waves, which are its familiar and prominent aspect, behaved in the familiar deterministic manner while the photons, which misbehave, were accepted only slowly as real "things." With electrons the situation is just the reverse; here it is the familiar particles that fail to conform. From the time of their discovery electrons had been thought of as bits of real material substance, having a well-known mass, occupying a definite point in space at any given moment and moving from point to point along well-defined paths.

But the interpretation of an experiment such as that of Davisson and Germer requires a profound modification in this deterministic, particlelike picture. To make this quite clear, it will be assumed that such an experiment is carried out by observing the angle of reflection for a large number of electrons one by one in succession. It will be found that there are certain discrete angles at which many have been reflected while few or none have come away at all others. As each electron goes to contribute its share to this overall distribution pattern, it seems to "know" about the arrangement and the spacing of the atoms through an extended region of the crystal lattice. A rational explanation for this behavior involves the concept of the de Broglie matter waves; these must therefore be recognized as a necessary concomitant of the electrons. Thus, like photons, electrons acquire a "dual" nature.

In optics the dichotomy of waves and particles was disposed of by showing that light is a well-defined single entity which has some but not all of the properties associated with the traditional concept of "waves" and also of "material particles." But electrons, or more generally material particles, *are* material particles. Here the problem of apparent "duality" arises when the concept of "particle" is given all of its classical connotations and no others. It is resolved by observing that the entity which really exists under that name is in some ways "like a particle" in that all the effects of its momentum and its energy are manifested at a single point in space; and in other ways it is "like a wave" in that its motion is affected by conditions extending in the manner of waves over a wider region. This is a logical and consistent observation concerning the nature and the behavior of bits of material substance; it is most decidedly not equivalent to the confusing assertion that these bits are in some inexplicable manner "both particles and waves."

The restriction on the sway of causality in the realm of mechanics is of such great consequences because of the fundamental importance of this discipline. Mechanics purposes to deal with all motions of all things under all possible conditions and for all possible reasons. This includes events occurring in living organisms as well as in the inanimate world. In particular, the principles of mechanics extend even to psychological phenomena so far as they may be related to physiological origins. Therefore, profound changes in the basic principles of mechanics may well have consequences in matters pertaining to enterprises of the human intellect, to metaphysics, ethics and theology (to be discussed later, particularly in chapters 15 and 18).

WAVES AND ATOMS

LOUIS DE BROGLIE was temperamentally disposed to pursue the broad implications of his novel ideas rather than to concern himself with applications to specific scientific problems. It is the Austrian theoretical physicist Erwin Schrödinger (1887-1961) who receives the credit for using the concept of matter waves to make a renewed attack on the problems of atomic physics. Based on the properties of these waves, he developed in 1926 a new system of physical concept and mathematical procedures which was given the name of *wave mechanics.*

A year earlier, and independently of de Broglie's work, the young German physicist Werner Heisenberg had made an entirely different approach to atomic problems. After receiving the Dr. Phil. at Munich under the guidance of Professor Arnold Sommerfeld, he went to work with Niels Bohr and his associates at Copenhagen. Through his intimate acquaintance with the ideas current in this group, he became convinced that the difficulties being encountered in atomic energy arose from the use of conceptions, such as electron orbits and transitions between them, which are impossible to observe experimentally. He felt that dealing with imagined models, which he referred to deprecatingly as "playing with paper dolls," has no place in science; and he set himself the task of building an atomic theory based entirely on facts of observation. Because he made use of an abstract mathematical technique call matrix algebra, his theory came to be known as *matrix mechanics.*

A matrix in the mathematical sense is an array of symbols arranged in rows and columns. Heisenberg's matrix, as shown in Figure 15, extends indefinitely to the right and downward. The

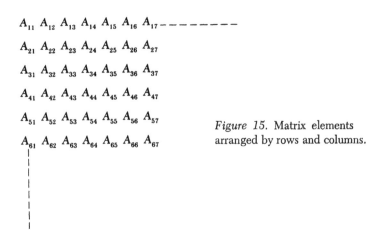

$A_{11}\ A_{12}\ A_{13}\ A_{14}\ A_{15}\ A_{16}\ A_{17} - - - - - - - -$

$A_{21}\ A_{22}\ A_{23}\ A_{24}\ A_{25}\ A_{26}\ A_{27}$

$A_{31}\ A_{32}\ A_{33}\ A_{34}\ A_{35}\ A_{36}\ A_{37}$

$A_{41}\ A_{42}\ A_{43}\ A_{44}\ A_{45}\ A_{46}\ A_{47}$

$A_{51}\ A_{52}\ A_{53}\ A_{54}\ A_{55}\ A_{56}\ A_{57}$

$A_{61}\ A_{62}\ A_{63}\ A_{64}\ A_{65}\ A_{66}\ A_{67}$

Figure 15. Matrix elements arranged by rows and columns.

symbols along the diagonal, with two like indices, are related to the sequence of quantized states of the atom having energies E, E_2, E_3, etc. These determine the energy of the photons, and the wavelengths or colors of the light, emitted by the atom. The symbols with unlike indices relate to the transitions between states and account for the number of photons, for the brightness of light, emitted at the various wavelengths. Thus, the matrix, as a whole, contains all that is observable in the spectroscopic behavior of the atom. For hydrogen and a few other simple cases the mathematical matrix calculations give results in excellent agreement with experimental observations. Yet there is no thought of trying to imagine how an atom is constructed or what goes on inside.

Schrödinger, on the other hand, preferred a model of the atom that could be visualized intuitively, a model built up of de Broglie matter waves. To him these were as real as light waves had seemed to physicists of the nineteenth century. Surprisingly, however, it was found that the two procedures developed by Heisenberg and by Schrödinger, though utterly different in conception, are nevertheless completely equivalent in results, that they are but two different mathematical techniques representing the same physical phenomena. In particular, both account naturally and consistently for that most characteristic attribute of atomic mechanics, the quantization of energy and of various other properties. Therefore, both new theories were brought together under the common designation of *quantum mechanics*, a highly appropriate name that has come into common use.

It would seem that Schrödinger's wave concept is quite unsuited to represent atoms, since it is the nature of waves to *travel*. There is, however, a manner of combining traveling waves to produce *standing waves*.

Standing waves may be set up, for example, on a stretched rope, as shown in Figure 16. If one end is tied to a post and the other end, held in the hand, is given a quick up-and-down flip, a hump is set up in the rope, at (1). It travels down to the post, is reflected there in reversed position and travels back to the hand. By sending a train of such humps down the rope, timed properly, it is possible to form the various patterns of oscillation shown at (2), (3) and (4). If the hand is moved at a low rate, the rope oscillates in a single loop, at twice the tempo there are two loops, at a triple tempo there are three, and so on.

Figure 16. Waves on a rope, showing a traveling wave and standing waves.

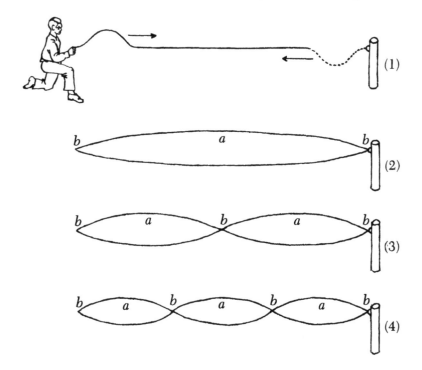

The superposition of two waves, the direct and the reflected one traveling in opposite directions, produces the standing wave. It is this *pattern*, and not the rope, which stands still. That is, the points marked *a*, at which the motion is widest, remain at the same place on the rope and so do the ones marked *b*, the *nodes*, at which there is no motion at all. (For a real rope the hand must be moved up and down slightly to maintain the motion. But an "ideal" rope, free of friction, would when once started keep on swinging without further input of motion, so that there would be a symmetrical pattern with a node at both ends.)

Standing waves are of common occurrence. Inside an organ pipe when it is blown there is a standing sound wave, the result of two waves, one traveling upward and the other downward after reflection at the upper end. More complex standing waves are formed in the sounding board of a piano or on the skin of a kettle drum.

Generally, standing waves can occur only in a situation where the waves are *confined*, as between the ends of a rope or between the top and bottom of an organ pipe. Furthermore, the standing waves must *fit* this limited region. The rope can perform sustained oscillations only in patterns which divide its length evenly into one or more whole loops; no others are possible. Similarly, the pattern of standing waves in an organ pipe must match the length of the pipe, a longer one having longer waves and a lower frequency or pitch. Since this behavior is common to all sorts of waves, it ought to be found as well with the de Broglie matter waves.

If it were true that the matter waves associated with the electrons in an atom form a pattern of standing waves, this pattern should be restricted to a sequence of certain particular forms. Furthermore, the transitions from one of these forms to another should happen discontinuously, since a gradual transition through impossible intermediate ones cannot occur. This is highly indicative of the discrete and discontinuous behavior observed in atomic phenomena, the one characteristic which most stubbornly defied all attempts at an explanation in terms of classical laws. Yet no one could have seen the relation between standing wave patterns and discrete atomic energy levels before de Broglie brought waves into the basic concepts of mechanics.

By considering certain analogies in the mathematical methods that had been developed for dealing with motions in complex mechanical systems and with the propagation of waves, Schrödinger built up a logical and consistent mathematical analysis of matter waves. He developed the *Schrödinger wave equation,* which, together with certain rules about the general behavior of these waves, takes the place of the classical laws of mechanics.

The laws of this new mechanics are quite as deterministic as the classical laws. Thus, matter waves do behave in a thoroughly predictable manner; it is the motions of the material objects guided by these waves which are not individually determined. Unlike the laws of classical mechanics, which deal directly with the positions and velocities of bodies, Schrödinger's equation is concerned with a quantity symbolized by ψ (psi), which represents the *magnitude* of the matter waves, varying from point to point in space and from moment to moment in time.

Schrödinger's laws show that the matter waves associated with atomic electrons do indeed form a pattern of standing waves which, through the action on the electrons of the attracting nuclear force, is restricted almost entirely to a very small region of space around the nucleus. Although the waves are not confined by an actual boundary, they do have the property of being limited to a sequence of discrete patterns.

In the application to any one kind of atom certain of its known characteristics must be entered into the Schrödinger equation. The solution then gives the particular sequence of discrete wave patterns for that atom, and with each of these states it associates a particular value for the energy. In addition, the solution yields information about the nature of the transitions between states. A transition which takes place more readily, gives rise to the emission of a greater number of photons and thus causes a greater intensity for light having the wavelength associated with that transition.

Since the pattern of matter waves representing the atom does not end abruptly but trails off gradually with increasing distance from the nucleus, the Schrödinger atom model has no definite *size.* Nevertheless, for atoms in the normal state of lowest energy, most of the wave pattern lies in a region whose size is in agreement with various experimental determinations of atomic diam-

eters. This result is gratifying. Schrödinger's calculations would
have been challenged if they had given a size differing markedly
from established values.

The true superiority of Schrödinger's conception lies, however,
in its *logical consistency*. All results follow naturally from the
basic principles; nowhere is it necessary to introduce special *ad
hoc* assumptions.

The sequence of natural numbers which came into atomic
theory through the quantization condition that Bohr used to se-
lect out the discrete sequence of electron orbits for the hydrogen
atom (page 57) appear naturally in the mathematical expressions for the sequence of wave patterns in Schrödinger's theory.
Thus, the numbers, 1, 2, 3, 4 . . ., which constitute the most
primitive example of a discontinous sequence, survive in the new
theory while the detailed apparatus of electron orbits is swept
aside.

For the hydrogen atom, in particular, the Schrödinger equation
must be "told" merely that it is dealing with a system consisting
of one electron and one proton. The resulting form of the equation shows that for the state of lowest energy E_1 the wave pattern
has appreciable magnitude only within a sphere of about 10^{-8}
centimeter in diameter, just the diameter of the smallest Bohr
orbit. For states of successively higher energy, E_2, E_3, etc., the
pattern extends out farther from the nucleus, again in the manner
of the Bohr orbits. The energy values are the same as those of the
Bohr theory and are thus not in precise agreement with spectroscopic measurements.

The entire scheme of the *chemical* properties of atoms, displayed in the periodic table of elements, has been related to an
arrangment of the atomic electrons in a sequence of groups (page
37) which Bohr visualized as a structure of shells, one outside
the other, containing successively a maximum of 2, 8, 18, 32, etc.,
electrons. While this scheme accounts for a great many experimental observations, it is entirely empirical, designed to conform
with the facts of experiment.

In the quantum-mechanical theory of atoms the numbers of
electrons in the various groups follow naturally from a general
property of electron waves, known as the *Pauli exclusion principle*, formulated by the Austrian theoretical physicist Wolfgang

Pauli (1900-1958), another of Arnold Sommerfeld's illustrious students. This principle asserts that no two identical electron wave patterns can exist in the same atom. Since it is found that there are only two variants of the wave pattern associated with electrons in the lowest energy level, there can be only two electrons in this state. For the next energy level there are eight possible variations; therefore, eight electrons may be assigned to it, and so on. The wave patterns of all the electrons in one group are approximately alike in size, and the energies associated with them are roughly alike as well. Strictly speaking, this enumeration of the possible distinct wave patterns in the various energy levels refers to those associated with the states of a single electron considered by itself. When many electrons are present in one atom, their identity becomes blurred and their wave patterns merge into a single one associated with all of them in common. Nevertheless, the numbering based on the Pauli principle is still valid. Thus, quantum mechanics supplies a sound theoretical support for the experimental observations of chemistry.

There is, moreover, another area of chemistry in which the *forms* of the wave patterns are of direct significance. Organic chemists, who deal with the many complex molecules built around carbon atoms, have found it necssary to devise *structural formulas* for these molecules, showing the arrangement of their constituent atoms. A rather simple formula of this kind, for the molecule of grain alcohol, is shown in Figure 17, the letters indicating the positions of carbon, hydrogen and oxygen atoms. Biochemists, who work with the exceedingly complex molecules of living organisms, some of which are built up of hundreds and even thousands of atoms, make use of models built of balls of different sizes and colors to represent the various atoms. Guided

Figure 17. The structural chemical formula of grain alcohol.

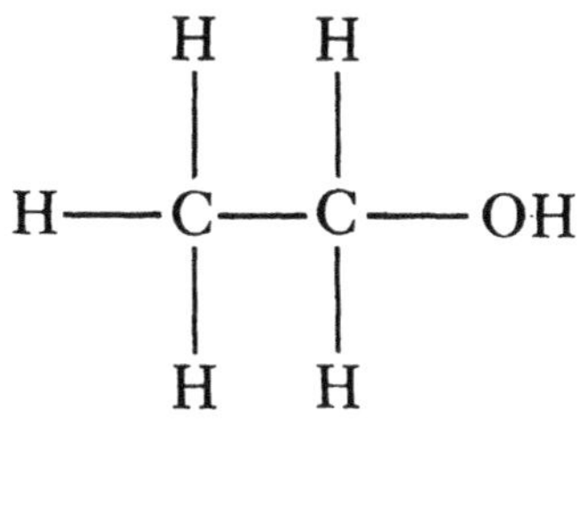

$$\begin{array}{ccc}
 & H & H \\
 & | & | \\
H - & C - C & - OH \\
 & | & | \\
 & H & H
\end{array}$$

by such models, chemists can put together and take apart these complex molecules and can attach or remove atoms singly or in groups at various selected locations. The success of such experimental procedures affords ample evidence for the reality of these structures.

There is nothing about the Bohr atom model, with its electrons whirling about in quantized orbits, to account for any of this. The quantum-mechanical picture of atoms does, however, afford a variety of symmetrical geometric wave patterns, some spherical, others more complex, that are well suited to serve as "building blocks" for these molecular structures. Exact mathematical calculations based on the Schrödinger wave equation are not possible for these intricate systems, but an approximate analysis has in many instances provided useful information about the ways in which such complex molecules are formed.

In the view of quantum mechanics a chemical synthesis, the linking of atoms or groups of atoms to form a new molecule, involves the overlapping and partial merging of their wave patterns. A general principle governs these reactions: if the new wave pattern of the combined molecule represents a more stable state than that of the separated parts, the reaction will occur; otherwise it will not. This accounts for the highly specific nature of chemical reactions, why hydrogen combines readily with chlorine but not with sodium, why an oxygen atom combines with just two of hydrogen, never with three, and so on.

Considerations such as these seem to lend a sense of reality to the matter waves. Indeed, in his work on the hydrogen atom Schrödinger was able to maintain a lively feeling for the real physical existence of his waves; but in the study of atoms with more than one electron this idea proved untenable. While the wave pattern of a single electron can be thought of as existing in three-dimensional space (so called because it has length, width and depth), the pattern for a two-electron atom must dwell in a space of six dimensions, that for three electrons in a space of nine dimensions, and so on. There is no possibility or need of visualizing these multidimensional spaces intuitively; they are purely mathematical constructs required by the nature of the Schrödinger wave equation appropriate to these situations. It seems highly dubious, however, to assign physical reality to a concept existing in a wholly mathematical sort of space.

This dilemma was resolved by a suggestion put forth by the German physicist Max Born, a leader in the mathematical development of quantum mechanics. Together with Schrödinger, who had just been awarded the Nobel Prize for his development of wave mechanics, he left Germany when Hitler came to power and settled in England. Later he too received the Nobel Prize for his contributions to the advance of theoretical physics.

By analogy to optics, where the calculated *intensity* of the light at a given place is taken as a measure of the *probability* of finding a photon at that place, Born suggested a similar probability interpretation of matter waves: the value of ψ^2 (not of ψ) in a given small region of space gives the probability of a particle's being in that region. (The *square* of the matter wave amplitude, rather than the amplitude itself, is chosen by analogy to light, where the intensity is proportional to the square of the light wave amplitude.)

For the hydrogen atom in the normal state in which the matter wave, found by Schrödinger's equation, is strong close to the nucleus and drops off rapidly at more remote points, Born's interpretation implies that for a large number of atoms considered as a group, most of them will have the electron near the nucleus. In a single atom the electron spends most of its time near the nucleus and is seldom farther away.

This conception does not do away with the multidimensional spaces for the matter waves. Since they are merely a mathematical measure of probability, they may be quite at home in abstract mathematical space. In this probability interpretation the electrons are still thought of as existing within the atom in the form of tiny, distinct particles. Schrödinger had pictured the electrons as being actually spread out over their wave pattern in the form of a tenuous cloud. The two conceptions do not differ, however, in their effect on actual phenomena, for the motions of the electrons within the atom are so exceedingly rapid that the far slower activities of both spectroscopic and chemical phenomena are determined only by their blurred time average distribution, as if they were "clouds" of electric charge.

The conception of matter waves as measures of probability is in accord with the previous discussion of causality in wave mechanics. While the behavior of the waves is deterministic and predictable by the laws of wave mechanics, these waves, in turn, govern only the overall statistical behavior of electrons, leaving their individual behavior quite unpredictable.

Lest it be thought that the analogy between standing wave patterns in the large-scale world and those of matter waves is quite complete, some important differences must be emphasized. With standing waves, such as those on ropes and in organ pipes, the energy depends on the *amplitude*. With matter waves the energy is related to the *form* of the standing wave pattern, those for states of higher energy being, in general, more complex. The amplitude of the waves is determined by quite different considerations. Since ψ^2, the square of the wave magnitude at a given place, determines the probability that a particle will be at that place, it follows that the summation of ψ^2 over all of space must be unity. The meaning of this may be seen by considering the tossing of a coin. The probability of tossing a head is one-half, and so is that of tossing a tail. The sum of the two, which is unity, stands for the probability of tossing either one, which is a certainty. To say that the sum of ψ^2 taken over all of space is *one* means simply that the particle is certain to be somewhere. With the sum fixed in this way, the value of ψ^2 and of ψ is determined at all points over a given wave pattern.

It might seem regrettable that the very evident advantages of the quantum-mechanical atom model have had to be achieved at the cost of relinquishing the neat and orderly orbits of the Bohr model. But in terms of the classical laws, if they be used consistently, this pretty picture is quite impossible. The conception which has been advanced of the atom as a tiny solar system, with the nucleus as the sun and the electrons as the planets, is highly misleading. In the solar system the mass of the sun is far greater than that of the planets and its gravitational force dominates all the motions, the weak forces between planets causing hardly perceptible perturbations. In the atom the forces between electrons would make shambles of the neat orbits in less than a billionth of a second. It surely must be reckoned as a gain for science that these impossible conceptions have been done away with.

After this discussion of the basic concepts of quantum mechanics it is next in order to consider certain remarkable general issues, and then a few specific applications in which this new mechanics has succeeded in shedding new light on old problems.

7

The Principle of Uncertainty

WAVE PACKETS

THE development of quantum mechanics has brought about profound changes in the scientific picture of the physical universe. The idea that objects move along well-defined paths in a predictable manner, according to deterministic laws, the idea that Newton had used so successfully in building a celestial mechanics, had to be relinquished. It was replaced by a new mechanics which declares that all motions are controlled by a mysterious sort of wave having but an abstract mathematical existence and exerting its influence only in an overall statistical manner. While this leaves the mechanical behavior of objects in the familiar large-scale world unchanged to all appearances, it depicts all things in the tiny world of atoms as patterns of waves, varying in space and time, within which the details of position and motion are undefined.

It is by no means true that all physicists were satisfied by this picture. Some insisted that any theory which falls short of a completely deterministic description of natural phenomena is necessarily incomplete and unsatisfactory. Remarkably, these dissenters included Albert Einstein, who initiated all the difficulties of noncausality by bringing particles into optics (and who expressed his dissent in the famous remark: "I cannot believe that God is playing a game of dice with the world!"), and Louis de Broglie, who augmented the difficulty by bringing waves into mechanics. Most physicists were willing, however, to set aside metaphysical problems and to take advantage of the obvious superiority of quantum mechanics in dealing with atomic phenomena. Heisenberg among others saw it as an added advantage that the new mechanics made it unnecessary to concern oneself with details that cannot be observed experimentally in any case.

This thesis could be challenged by suggesting that someone might actually succeed by new and refined experimental procedures to observe and measure the precise positions and motions of electrons within the atoms. Heisenberg faced this issue and found that the possibility of making such observations is not a matter of improvement in experimental techniques. He succeeded in showing that the basic principles of quantum mechanics prohibit these precise measurements, even in principle.

To show how this bold conclusion may be justified, it is appropriate to consider, not an electron in an atom, but a much simpler situation: a single material particle free to move by itself through space. To fix ideas, it will be assumed that it moves along a straight line extending indefinitely to the right and left. The associated matter waves will also be traveling along this line. Since they must exist wherever there is a probability of finding the particle, they too must extend over all of its infinite length because a priori the particle may be anywhere on the line. When the moving particle has been observed experimentally at a given instant, its position at that instant is fixed more or less precisely near some particular point on the line. Its associated matter waves must then also be restricted to a short segment of the line. The problem is to show how it is possible to construct a short piece of wave, a traveling *wave packet*, out of a group of infinitely extended traveling waves. This requires a rather lengthy discussion which is justified by the importance of the conclusions deriving from it.

Assume that two waves of the same amplitude traveling along the line differ in wavelength, 100 of the longer waves equaling 101 of the shorter. At successive points separated by 100 wavelengths the two waves will be in phase, waving up and down together, and will add up to a strong composite wave. At points midway between, where the longer wave has gone through 50 and the shorter through 50½ wavelengths, the two are opposite in phase and add up to zero. If the wavelengths are in the ratio of 1000 to 1001, the points at which the waves are in phase will be separated by 1000 wavelengths; and if the difference in wavelength is infinitesimally small, there will be but one such point (the others being infinitely far away). There is a considerable composite amplitude at other places along the line. If, however, a very large number of waves are combined, differing successively in wavelength by an infinitesimal amount, there will be one small region in which all of them are very nearly in phase. Here the composite wave has a large amplitude while at all other places it is relatively negligible. An exact mathematical analysis of such a group of infinitely extended de Broglie waves shows that a wave packet of limited extent can be formed in this manner, and that it travels along the line at just the velocity of the particle with which the waves are associated.

The significance of the wave packet is made evident by considering the relation of its *extension* along the line and the *range of wavelengths* of which it is constructed. If this range is small, if the longest and shortest of the waves making up the packet differ only a little in wavelength, the packet extends for a considerable distance either side of its maximum point, as shown in figure 18 at (1). If the packet includes a greater range of wavelengths, it is narrower, as seen in the figure at (2).

Figure 18. Diagram of a wide and a narrow wave packet.

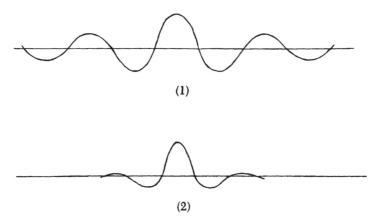

(1)

(2)

From the general conception of guiding matter waves, it follows that the position of the particle is fixed only approximately by the requirement that it must be somewhere within the bounds of the traveling wave packet. Because of the de Broglie relation between wavelength and momentum, the momentum of the particle may have a range of numerical values corresponding to the range of wavelengths contained in the packet. This leads to a new and important conclusion concerning the precision with which the position and momentum of a particle may be determined. If the wave packet is short and the position is fixed within narrow limits, its momentum may have a wide range of values since the packet includes a wide range of wavelengths. Conversely, if it is made up of a narrow range of wavelengths so that the momentum is rather precisely determined, the position is fixed only within a wide range.

Two extreme situations are possible. If the range of wavelengths includes all sizes from zero to an infinite length, the packet shrinks to zero width and the position is determined with ideal precision while the momentum may have any value whatsoever. If, on the contrary, there is only one sharply defined wavelength, the momentum is precisely determined; but this single wave extends at constant amplitude over the whole infinite length of the line, and the particle may be anywhere.

Heisenberg spoke of his new conception as the *Unschärfeprinzip*, which translated literally is *unsharpness principle*, in other words, the principle of limited precision. Later he used the somewhat ambiguous term *Unsicherheitsrelation*, which has been translated as the *Principle of Uncertainty*, a name that has come into general use.

It is important to note that the uncertainty principle does not set limits to the precision with which either the position alone or the momentum alone may be determined. But it does show that there is an ultimate limit to the precision with which *both* may be determined simultaneously. This limit is not set by practical considerations of precision in the construction and use of scientific instruments. It lies in the very nature of things. Here it was seen to follow inevitably from the concept of matter waves. Heisenberg showed that it results just as directly from the mathematical principles of matrix mechanics.

The formula which Heisenberg derived for this ultimate limit of precision, for the minimum uncertainty with which position and momentum can be determined simultaneously, is:

$$\Delta p \times \Delta q \geqslant h/2\pi$$

Here p is, as before, the symbol for momentum while q designates the position. The letter Δ (delta), which is a general mathematical symbol for "a small change in," here means "the uncertainty of." The symbol connecting the two parts of the formula means "equal to or greater than." Thus, the formula reads: the product of the uncertainties in momentum and in position is equal to or greater than (is at least equal to) Planck's constant divided by 2π. The formula shows again that there is no limit to the precision of either p alone or q alone, for either Δp or Δq may be made as small as desired. But as one becomes smaller, the other must increase to maintain the product unchanged. This is in accord with the results derived from the discussion of wave packets. Furthermore, the formula sets a *lower* limit to the uncertainties. Simultaneous determination of position and momentum cannot possibly be *more* precise, even in principle, but they can of course be less so.

PRECISION OF MEASUREMENT

IN his publication of 1927 in which he first announced his new principle, Heisenberg felt constrained to make it more acceptable by giving examples of ways in which simultaneous measurements of a particle's position and momentum might actually be carried out. He deliberately chose to discuss "ideal" experiments, unhampered by practical limitations on precision. He suggested that a single electron might be observed as it passes through the field of view of an ideal microscope capable of working with "light" of any wavelength. If the electron e is to be "seen," at least one photon $(h\nu)$ must collide with it, as shown in Figure 19, and rebound into the microscope. Since the photon is guided by waves, it does not necessarily proceed in a straight path from the electron to the eye; rather, because of diffraction it is deviated in an undetermined manner as it passes into the microscope. It is impossible to trace its path precisely back to the electron, whose position at the moment of the collision is therefore in doubt. In the

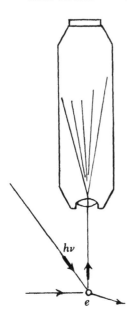

Figure 19. Observation of an electron by an ideal microscope.

collision the photon's impact deflects the electron and changes its momentum in a manner that cannot be precisely known since the photon's angle of rebound is not known. The uncertainty in position can be minimized by using photons of shorter wavelength, for which there is less deviation due to diffraction. But shorter waves imply photons of greater energy, which disturb the electron more strongly in the collision and thus cause greater uncertainty in its momentum. Conversely, less energetic photons permit more precise determination of the momentum but render the knowledge of the position less precise.

Clearly, the uncertainty principle is closely related to the quantization of radiant energy in the form of photons. If this energy were carried by the light waves in the manner of the classical theory of wave propagation, the intensity could be reduced, and also the wavelength could be shortened, without limit. Therefore, both position and momentum could be determined, in principle, with ideal precision. Such considerations show that the uncertainty principle has its roots in the very foundations of quantum mechanics.

Nevertheless, this principle was not easily accepted. Many ingenious "thought experiments" were devised (some of them by Albert Einstein) purporting to circumvent the limitations it imposes on ultimately attainable precision. But these were invariably countered by equally ingenious ones in its support. Gradually, uncertainty in the sense proposed by Heisenberg came to be recognized as a basic principle in science.

Again the question of why a concept of such fundamental significance was not discovered in the world of everyday affairs finds its answer in the extremely small numerical value of Planck's constant. If the momentum of a one-gram bullet could be determined to within 10^{-12}, corresponding to a measurement of the velocity to a precision of one-millionth-millionth centimeter per second, the determination of its position could not possibly be better than to within one-thousandth-millionth-millionth(10^{-15}) centimeter. Such measurements are obviously impossible; the practical limitations on attainable precision are incomparably greater than the ultimate limits set by Heisenberg's principle. With atoms the situation is far different. Even if the position of an atomic electron is known no more precisely than that it is somewhere within the atom (within about 10^{-8} centimeter), the corresponding uncertainty in the momentum is such that the velocity cannot be known more precisely than within about 10^8 centimeters per second. But this is approximately the value of the entire velocity of an atomic electron. The uncertainty in velocity is thus equal to the velocity itself; in other words, the velocity is quite undetermined. This is Heisenberg's answer: it lies in the very nature of things that the details of motion of atomic electrons can never be observed.

There are situations in which both the position and the momentum of a single electron, and even of a single photon, may be determined simultaneously with a precision that is acceptable as a good measurement, well within the usual limits of experimental error. This occurs when these particles are moving in a large system, as is the case in the Compton effect experiment (page 54). Here the uncertainty formula shows that, with the range of wavelengths in the wave packet limited to fix the momentum to within one-tenth percent, the uncertainty of position, or the spread of the wave packet, is only about one-thousandth centimeter. For an

X-ray photon a similar uncertainty of its momentum (that is, of its wavelength) gives an even smaller uncertainty in its position. These wave packets are so small in relation to the whole apparatus that the photons and electrons may be treated as particles obeying classical laws of mechanics, as was actually done in the discussion of the Compton effect.

The uncertainty principle has sometimes been misconstrued to imply that a particle actually *has* both a precise position and momentum until it is disturbed by the experimenter, that the act of observing the position precisely "destroys" the precise momentum; in other words, that nature is involved in a bizarre conspiracy to prevent the discovery of something that has real existence. Contrary as it may seem to all familiar experience, it is nearer to the truth to assert that a particle of itself has neither a position nor a momentum and that the act of observation *creates* its mechanical state. This conclusion follows directly from the general quantum-mechanical principle that all of the circumstances of the particle's position and momentum are contained in its accompanying matter waves.

In the example of the single particle free to move through space, it was seen that its wave pattern could assume various forms between the extremes of a single wave, extending uniformly over all of space, and a sharply concentrated pulse composed of an infinite number of such waves. All of these are potentially possible, but which of them is actually realized depends on how the particle is observed. If the position is fixed closely, the wave packet is highly concentrated; if it is the momentum that is closely determined, the packet is more extended. Thus, it is not the particle itself but the particle and the observing apparatus which *jointly* exhibit particular values of position and momentum.

This reasoning applies quite generally. A given situation is not simply "there," as in the classical conception, fully determined and ready to be investigated in detail. Rather, it is the joint existence of the situation to be studied and the techniques of observation which together represent more or less precisely its specific mechanical properties.

Niels Bohr gave much thought to the meaning of the Heisenberg uncertainty principle and its implications for the relation between observed object and the observer. By careful reasoning he was able to show that any apparatus designed to measure position with ideal precision cannot provide any information about momentum, and vice versa. Thus, two mutually exclusive experiments are needed to obtain full information about the mechanical state, each complementing the other. He expressd this conclusion as a general *principle of complementarity*.

The idea of complementarity is closely related to the wave-particle "duality." In a situation such as in the experiment of Davisson and Germer (page 74) the observation of the discrete angles of strong reflection affords a precise determination of the electron's wavelength, that is, of its momentum. But the position is determined only to the extent that the electron is known to be somewhere in the apparatus. Here the waves are spread out over a wide area and manifest their presence clearly. In an experiment in which the position is closely determined, the wave packet is concentrated in a particlelike manner, and it no longer shows wavelike behavior. The Heisenberg principle, which asserts that both of these aspects cannot be observed in one and the same experiment, is in accord with Bohr's conclusion concerning mutually exclusive experiments. It also shows that wavelike and particlelike properties can never come into conflict experimentally, a conclusion reached previously (page 65) on other grounds.

The uncertainty principle applies to another important pair of quantities: *energy* and *time*. Here it takes the form:

$$\Delta E \times \Delta t \geqslant h/2\pi$$

This states that the product of the uncertainty in the energy which a mechanical system possesses and the uncertainty of the time at which it has this energy is at least equal to $h/2\pi$. Here again the small value of h assures that this uncertainty is of no consequence in the mechanical behavior of large bodies. Applied to atoms, however, it does dispose of attempts to describe in detail the transitions between energy states, whether envisioned as jumps between Bohr orbits or Schrödinger matter wave patterns. Because of the high precision of spectroscopic measurements of wavelength and frequency and therefore of energy levels, the question of how the transition proceeds in time is impossible to answer by experimental observations and is therefore meaningless.

That the laws of quantum mechanics are not deterministic with regard to the mechanical behavior of individual objects is familiar. The uncertainty principle shows clearly how this comes about. Considering again a single particle free to move along a line, it is evident that the precise *prediction* of its position at some instant in the future requires the precise knowledge of *both* its position and its velocity at the present moment. This, the uncertainty principle asserts, is impossible. Thus, the impossibility of predicting future events precisely is not due to an inadequacy in the method of prediction, but to the impossibility of obtaining the precise knowledge of the *initial conditions* upon which the prediction must be based.

The process of prediction consists, in the language of wave mechanics, in describing in detail the future course and development of a given wave packet; and this is always possible since matter waves are subject to rigorously deterministic laws. These laws state that matter waves of different wavelengths do not travel at the same velocity. This general rule makes precise prediction of position impossible in any case, for in general a wave packet contains waves of various wavelength. Therefore, as it travels along, some of the waves move ahead faster, some lag behind, and the packet spreads out. If it is initially highly compact and is made up of a broad range of wavelengths, it spreads out rapidly, thus quickly increasing the uncertainty of position. On the other hand, if it is made up of waves varying over only a small range of wavelength, it spreads out less rapidly but is, of course, broader to begin with. In either case prediction of future position is in doubt and becomes more so as time goes on.

If the particle is free of forces, the wavelengths in its wave packet do not change, and the uncertainty in its momentum remains the same. If forces are acting, however, the momentum and its uncertainty are, in general, altered. The future changes of the wave packet are thus predictable, and so are the changes in the *uncertainties* of position and momentum; but the changes in position and momentum themselves are not.

More specifically, the probability of finding the particle at any point within its wave packet is proportional to the square of the packet's amplitude at that point. In general, the various waves making up the packet have different amplitudes; and the probability that the particle may have the momentum corresponding

to the wavelength of any one of these waves is again proportional
to the square of the amplitude of that particular one.

FURTHER IMPLICATIONS OF QUANTUM MECHANICS

IN the whole discussion of how quantum mechanics developed, starting with Max Planck's discovery of energy quantization
in 1900, the importance of the constant h has been apparent. It is
interesting to observe how the whole nature of the physical universe is dependent upon the numerical value of this constant and
to speculate on how profoundly things would be changed if this
value were very different. If h were so small, so close to zero, as
to be of negligible effect in all possible situations, Planck could
not have detected any evidence of energy quantization in his
oscillators, nor could there be the quantized energy levels conceived by Bohr. De Broglie's matter waves would have an infinitesimal wavelength; there would be no diffraction effects, and no
departure from classical behavior, even in the tiniest of systems.
Heisenberg's uncertainty would be zero; and the deterministic
laws of classical mechanics would be universally valid, a highly
desirable state of affairs, so it would seem.

However, if Planck's constant were zero, there would have been
no Planck, and indeed no rational beings, or any forms of life, for
it is quantization that accounts for the existence of stability and
organization in the atomic substratum of the universe. Because
the energy content of atoms is restricted to certain discrete values
(page 57), an assault of considerable energy is needed to jolt
them out of their normal state, and afterward they return quickly
and precisely to normal. Without quantization there could be no
definite normal state. Any electronic configuration whatsoever
would be possible, and the slightest disturbance could alter this
configuration permanently. Atoms would have no stable and
specific properties. There would be no well-defined organization
of atoms into molecules or of molecules into large structures. The
universe would be a formless and meaningless blob without history, plan or purpose. In our present earthly environment quantization alone makes atoms act—to use Newton's words—like the
"solid, massy, hard, impenetrable particles" formed by God in the
beginning "that nature may be lasting."

The numerical value of Planck's constant, which tends to keep the atom world stable and well organized, also introduces discontinuity and indeterminism, but on such a minute scale that for things of substantial size all motions appear smooth and predictable. If this constant were a billion billion billion times greater, discontinuity and undeterministic behavior would obtrude conspicuously into affairs on the human scale. Nothing could be controlled or predicted; orderly conduct would be impossible. That rational beings could develop under such conditions would seem impossible.

Although scientists still use the deterministic laws of classical mechanics to deal with events among objects which can be readily seen and handled, their attitude toward the operation of causality in such events has been profoundly altered through the new insights afforded by quantum mechanics. They no longer believe that individual events happen the way they do because of the operation of rigorous deterministic laws. Rather, things are thought to happen the way they do because it is highly probable that they do so and highly improbable that they do otherwise.

The categorical "always" and "never" of classical laws have been replaced by "usually" and "seldom." Since the pronouncements of science should be based upon the evidence of experimental observation, these more cautious words are certainly to be preferred, because such evidence, no matter how extensive, can never be either universal or final.

Quantum mechanics has most assuredly provided a greatly enhanced comprehension of the world of atoms and has given an estimable account of all the observable phenomena arising from the atomic substratum. But it has blurred the details of this substratum; indeed, it has shown that they are unknowable even in principle. However, as sober reflection shows, these details had never been actually observed. They are mental constructs built of concepts borrowed from the world of familiar experiences, and they were never justified either experimentally or conceptually.

A poet might speak of a never-never land which, in its very nature, is forever beyond our ken, but such a concept does not belong among the working hypotheses of scientists. The development of quantum mechanics has been a sobering experience. It has shown where the limits of intuitive comprehension lie.

8

Applications and Conclusions

THE PARTICLE IN A BOX

THE principles and concepts that characterize the new quantum mechanics, and serve to distinguish it from nineteenth-century physics, may be made clearer and more meaningful by considering actual examples in which the peculiar results of quantum-mechanical analysis become apparent. One might, for example, discuss atomic phenomena where the electrons, "boxed in" by the nuclear attractive force, act in a characteristic quantized manner. But this "box" is not a simple one, and the mathematical analysis of the resulting standing wave pattern is somewhat complex. There is, however, a situation which is far simpler but does, nevertheless, exemplify typical quantum-mechanical results.

Consider a particle of mass m which, as shown in Figure 20, is free to move to and fro along a straight line in a "box" of length L. Barriers set up at the ends, indicated by the vertical lines, are thought of as being impenetrable so that it is impossible for the particle to pass beyond them. The energy of this particle is wholly energy of motion, known as *kinetic energy*, which is related to its mass m and velocity v according to the formula:

$$\text{Kinetic energy} = \tfrac{1}{2}mv^2$$

With the moving particle there is associated a pattern of standing waves whose form is found in the usual way by solving the

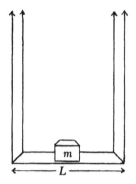

Figure 20. A mass particle m confined in a "box" of length L with impenetrable ends.

appropriate Schrödinger equation. This is set up by using simply the information that the particle has a mass m and is confined to a line of length L. The solution takes the form of a smooth wave, the *sinusoidal* wave shown in Figure 21.

The principles of quantum mechanics require that the waves "fit" the box, that the wave amplitude ψ be zero at both ends of L. This follows from one of the rules governing the behavior of matter waves: ψ may have only one value at one and the same point. Since ψ is necessarily zero at points outside the barriers, where the probability of finding the particle is nil, it must also be zero at the two points directly at the barriers. This leads immediately to the result that only certain particular wavelengths are possible, only those for which an integral number n of half-wave loops spans the length L, as seen in Figure 21 in the wave forms corresponding to the values of 1, 2, 3 for n. Any other wavelengths, such as those shown in the fourth section, are impossible.

Because of the de Broglie relation between wavelength and momentum ($\lambda = h/mv$), it follows that the momentum, and consequently the kinetic energy, is also restricted to a sequence of discrete values. Thus, the analysis of the motion in terms of matter waves leads naturally to the quantization of energy.

The solution of the Schrödinger equation gives, in addition to the wave patterns, the formula for the quantized energy values $E_1, E_2, E_3 \ldots$ thus:

$$E_n = \frac{n^2 h^2}{8mL^2} \qquad n = 1, 2, 3 \ldots$$

The lowest value (for $n = 1$) is therefore:

$$E_1 = \frac{h^2}{8mL^2}$$

and the succeeding values are just 4, 9, 16, etc., times greater. The significance of the relations expressed by these formulas will be made apparent directly.

It is instructive to calculate the numerical value of E_1 for a particle of electronic mass $m = 9.1 \times 10^{-28}$ gram confined to a box whose length is equal to the diameter of the hydrogen atom in its normal state ($L = 10^{-8}$ centimeter). This is found to be approximately the same as the actual energy of the hydrogen atom electron in its state of lowest energy. Since the electrons in the atom and in the box are in very different conditions, this rather surprising result shows to what extent quantum-mechanical analysis is independent of unobservable details; only the mass of the electron and the general size of the region in which it moves,

Figure 21. Sinusoidal waves, the general form of the wave-functions ψ for a mass particle in a box.

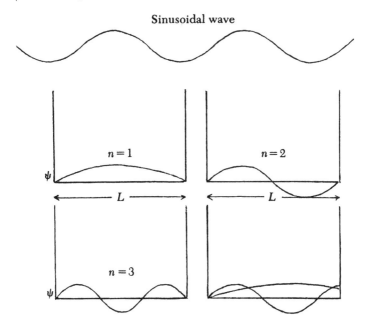

Sinusoidal wave

both observable quantities, are important. However, the higher energy values for the simple box and for the atom do differ greatly, for here the difference of condition becomes apparent. The atomic electron takes over a larger region of space at higher energies while the box remains unchanged in size.

The formula for the energy E_n shows that all the discrete energy values, and the steps between them, vary *inversely* as the magnitudes of m and of L, that is, these energy steps are *greater* for *lighter* particles of mass confined in *smaller* boxes. In the range of atomic magnitudes the energy steps are relatively large, and the discrete quantized values are clearly evident. But for a particle and box of substantial size the successive energy values lie too close to be distinguishable. Any motion that is at all observable corresponds to an energy state having a numerical value of n in the billions of billions, and the smallest observable change in motion means a sensibly smooth transition through billions of successive states. This is a specific example of the rule that for mechanical phenomena on a large scale quantum mechanics leads to results in agreement with the classical laws.

In particular, the formula for E_1 implies that the minimum energy can have the value zero only for a box of infinite length. This means that the particle cannot be at rest, with zero energy of motion, unless it is in free space; if it is confined to a region of finite length, it must necessarily be moving. This apparently absurd result is reconciled with familiar experience when numerical magnitudes are considered. It appears that for boxes and particles of visible size the motion in the state of minimum energy is so exceedingly slow (perhaps a hair's breadth in a century) as to be completely unobservable.

Conversely, the formula shows that, as the particle is confined to a smaller box, the minimum value E_1 which its energy can have becomes greater. This relation is found to be true as well for the electron in the hydrogen atom. Here also the least possible energy which the electron can have is greater when it is confined more closely to the nucleus. In a large orbit, corresponding to a state of higher energy, the electron has more energy than the minimum amount pertaining to this large region, so it may emit a photon and pass to a smaller orbit. In the normal orbit it has just the energy needed to maintain itself there. If it "tried" to move in still closer to the nucleus, it would find itself with less than the minimum energy needed to exist in this more confined state; therefore, it cannot possibly do so. In the language of de Broglie waves this means that the nuclear force is incapable of producing a stable wave pattern more concentrated than that of the normal state. This conclusion is valid generally for the innermost electrons of all atoms.

In this simple and direct manner quantum mechanics disposes of the problem of *atomic stability*, which had burdened atomic theory ever since Rutherford proposed the nuclear atom model.

The wave patterns of the particle in a box show a property that is found as well in atoms; those corresponding to higher energy values are more complex. In particular, they have more *nodes*, places at which the magnitude of the wave is zero. The pattern of lowest energy has just two nodes, one at each end of the box. For the second energy level there is an additional node at the center; for the third level there are four nodes in all.

In atoms the wave patterns are arranged in the three directions of space rather than along a line, so various kinds of nodes are possible. Some of these divide the pattern into concentric shells, one outside the other. Other nodes may separate it into two or more lobes, arranged to the right and left or in other symmetrical patterns about the nucleus. These nodes produce the various symmetrical wave patterns which were mentioned in the discussion of complex molecular structure (page 88).

In the characteristic manner of quantum mechanics the analysis of the particle in a box affords no detailed information about its position or motion. The magnitude of ψ^2 at any point gives only the probability that the particle will be at that point. Relative values of this probability, for the first, second and twentieth energy state, are shown in Figure 22. While the curves for ψ swing both above and below the center zero line to positive and negative values, ψ^2 is positive everywhere as is appropriate for a quantity that represents probability, since negative probability has no meaning.

The section of the figure for $n = 20$ indicates clearly that, for a large particle where a motion sufficient to be detectable corresponds to a state well above the billion billionth energy level, the probability is sensibly uniform over the length of the box. This is

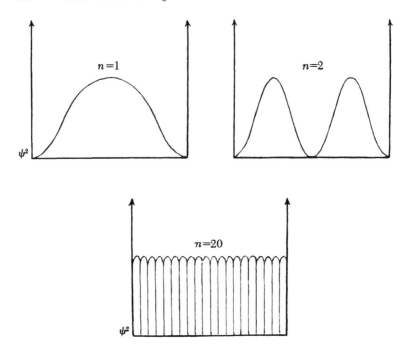

Figure 22. Patterns of ψ^2 for the first, second and twentieth energy states of a particle in a box.

in accord with the fact that a large, freely moving particle has a constant rate of to-and-fro motion through its path.

For a box of atomic size the details of motion are unobservable, as the Heisenberg uncertainty principle shows. If, however, the moving particle were an electron and could absorb and emit photons, quantum mechanics could account for its *observable* energy changes in detail.

NATURAL RADIOACTIVITY

A PARTICULARLY interesting example of how quantum mechanics has clarified a previously obscure subject is that of radioactivity. The naturally occurring emission of various kinds of rays from certain *radioactive* substances was first observed by the French scientist Antoine Henri Becquerel (1852-1908) in the closing years of the nineteenth century and was studied intensively by Marie and Pierre Curie shortly thereafter. Their work, together with that of several other eminent physicists, led to the recognition that there are three distinct kinds of rays, named *alpha, beta* and *gamma,* and that the source of all three is the atomic nucleus.

Alpha rays, or alpha particles as they were called later, were found to be tightly bound groups of two protons and two neutrons, identical with the nuclei of helium atoms, forming a subgroup within the larger group of protons and neutrons in the heavier nuclei. The emission of an alpha particle reduces the positive charge of the nucleus by two electronic units and thus causes the atom to be transformed into another kind that is two places lower in the periodic table of the elements. A radium atom (atomic number 88), whose nucleus contains 88 protons and 138 neutrons, is transformed by alpha emission to a substance called radon, whose nucleus has 86 protons and 136 neutrons.

Beta rays were found to be the electrons discovered by J. J. Thomson. There is good experimental evidence to show that beta emission occurs when one of the neutrons in the nucleus transforms itself spontaneously into a proton plus an electron (plus another very unusual kind of uncharged particle, the neutrino, to be discussed later). The electron leaves the nucleus immediately while the proton is retained. The atom thus acquires one additional proton in its nucleus and is transformed into a new atom having an atomic number greater by one unit. These processes are called radioactive transformation or *radioactive decay.*

Gamma rays are photons. They are always emitted subsequent to alpha or beta emission. The nuclei newly created by the emission of electrically charged particles are not necessarily in their normal, most stable state; and they quickly drop to this state by the emission of a photon, a process analogous to the dropping of an electron to an orbit of lower energy in the Bohr atom model. Since the forces between particles in the nucleus are very great, gamma-ray photons have very high frequencies.

The activity of alpha particles within the nucleus is related in some ways to that of the mass particle in the box. Their condition in the nucleus has been likened to that of a ball inside a well sur-

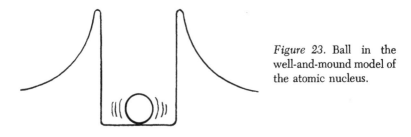

Figure 23. Ball in the well-and-mound model of the atomic nucleus.

rounded by a barrier in the form of a sloping mound (Figure 23). The walls of the well, though not an absolute barrier as were the sides of the box, normally keep the ball inside; but if it manages to get over the top, it promptly rolls away down the slope. Similarly, an alpha particle is held in the nucleus by the strong but short-range binding forces (to be discussed in chapter 12) acting between the nuclear particles. If it frees itself of these forces, it is pushed away by the electrical repulsion between its own positive charge and that of the nucleus, acting like the slope of the mound; the push is strong while it is very close and becomes more gentle as it moves farther away. The concept accounts for the observed high velocity of the ejected alpha particles. (The forces within and around the nucleus exist in all directions, right-left, up-down, fore-aft. The figure may be thought of as representing the condition along one particular line drawn through the nucleus.)

The well-and-mound model of the nucleus is a plausible one, but in terms of the classical laws of mechanics it leads to a serious difficulty. It would seem that an alpha particle having sufficient energy to get out of the nucleus would do so immediately while one that does not have this energy would never get out. Therefore, radioactive atoms should decay in a single tremendous flash of radioactivity, or there ought to be no radioactivity at all. Actually, many radioactive materials decay very leisurely. For radium the decay proceeds at such a rate that in a given sample one-half is transformed to radon in 1620 years, a period known as its *half-life*. Uranium decays far more slowly; its half-life is about four and one-half billion years, a period comparable to the estimated age of the earth.

Various attempts were made to explain these experimental observations in terms of classical concepts. It was assumed, for example, that the nuclear particles might be in violent random motion within the nucleus so that by chance one of them might acquire for an instant considerably more than the average amount of energy and thus manage to escape. This idea became untenable when various experimental findings indicated that there is within nuclei a well-defined and stable organization among the protons and neutrons. It was further suggested that nuclei might "age," going through a slow process of change that makes them gradually more prone to radioactive decay. In addition to being at odds with the well-substantiated general concept that all atoms of one kind are exactly alike at all times, this idea could not account for the manner in which radioactive decay is found to proceed.

To understand the observed mode of radioactivity, it is instructive to consider an analogy of mortality in a group of living beings. Assume that there exists a race of men who do not age or change in any way as the years go by, who are inherently immortal, and die only as the result of accidents. The probability of having a lethal accident is the same for all of them and does not change with time. This probability is such that of an original group of 1000 half will die in the following 50 years. Since the 500 survivors have not changed, half of them will again die in the following 50 years. Thus, at the end of successive 50-year intervals the number of survivors will be 500, 250, 125, and so on.

Among ordinary mortals the situation is far different. Of 1000 newborn men it might well be true that 500 will survive 50 years later, but it is highly improbable that 250 will reach the age of 100 years. It would be necessary to devise a very peculiar mode of aging to obtain results such as those due to purely accidental death.

All the experimental evidence indicates that radioactive decay is a process due to pure chance. No test can determine whether any one individual radium atom will decay today or thousands of years hence. In all radioactive materials the process of decay proceeds so that the fraction of the original sample remaining at the end of successive half-life periods is 1/2, 1/4, 1/8, 1/16, 1/32, etc. This is known as the *law of radioactivity*. It is evident that among any one kind of radioactive atoms the chance of decay is the same for all at all times. No reasonable alternative explanation has been advanced.

Probability is the familiar *modus operandi* of quantum mechanics, and it is reasonable to expect that it might provide an adequate explanation for radioactive decay. A quantum-mechanical analysis was carried out successfully, largely through the work in the United States of the Russian-born physicist George Gamow. He obtained information about the nature of the "mound" and the "well" of atomic nuclei from alpha particle bombardment experiments.

Such experiments are analogous to the procedure of rolling balls up against the mound-and-well model of the nucleus. A ball sent against the mound with moderate energy will roll up part of the slope and then "rebound"; but given enough energy, it will reach the top and drop into the well. Similarly, an alpha particle projected from without at an atomic nucleus will ordinarily rebound elastically; but if it approaches with sufficient velocity to make contact, it is drawn in by the nuclear binding forces which are amply strong enough to overcome the electrical repulsion. (The

Figure 24. Patterns of ψ waves, in the well-and-mound model of the atomic nucleus.

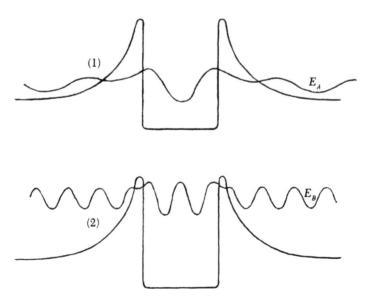

entrance of the alpha particle usually creates an unstable condition of the nucleus, which may revert to a stable state by the emission of one or more of its protons or neutrons. This process is called *artificial radioactivity* since it does not occur naturally in the environmental conditions on earth.)

Using information obtained from these experiments, Gamow succeeded in setting up and solving the Schrödinger equation appropriate for the kinds of forces involved in alpha-particle emission. He obtained patterns of matter waves as shown in Figure 24. (Here again the figure shows the wave pattern along one line only; actually it extends in all directions around the nucleus.) Inside the well the waves are much like those in the box with impenetrable walls, but the fact that the mound is not insurmountable has the consequence that the waves extend *beyond* the confines of the well. As they pass through the barrier, they are attenuated so that their amplitude outside is, in general, greatly reduced.

In accord with the usual interpretation of matter waves, this means that the particle has a *probability of being outside* even though it does not have sufficient energy in terms of classical concepts to get out.

Studies of the energy changes involved in nuclear phenomena show that in different kinds of radioactive atoms the alpha particles lie at various energy levels within the nucleus. For a nucleus of the kind shown at (1), where the alpha particle is at a level far below the top of the mound, the wave must penetrate a tall and broad barrier and is strongly attenuated. For one in which the particle lies at a higher level E_B, as at (2), the wave outside has a greater amplitude. Since there is in this instance a greater probability of radioactive emission, this kind of radioactive material has a shorter half-life. (For an infinitely high mound the attenuation of the wave is complete, and the wave amplitude outside is zero, as is the case for the box with impenetrable barriers.)

A well-known, but previously unexplained, law of radioactivity, the Geiger-Nuttall law, states that radioactive atoms having short half-lives emit alpha particles at high velocity. The well-and-mound model accounts readily for this relation. A particle situated at a high energy level as at E_B corresponds, as has just been shown, to a short half-life. After escaping, it finds itself very close to the nucleus, where it receives a strong push and flies off at a high velocity. But one leaving the nucleus at the low level E_A, for which the half-life is long, is pushed away more gently.

Half-lives of various radioactive substances vary over an enormous range, from billions of years to fractions of seconds. The former are much like the classical case of particles that never get out, the latter like those that get out immediately. Whereas classical mechanics gives only the extremes of "always" and "never," quantum mechanics can account for all the gradations between "seldom" and "frequently." (Substances with extremely short half-lives do not cause a tremendous blast of radioactivity. They come into being as the result of a previous, much slower radioactive process; and they cannot decay faster than they are formed.)

Most kinds of atoms have stable nuclei; they are not radioactive. Perhaps it is better to say that their half-lives are so extremely long, possibly billions of times the age of earth, that their radioactivity is too weak to be observable. For these atoms the alpha particles lie at deep energy levels, well below the level of the "plain" out beyond the mound.

Finally, it is appropriate to explain how half-lives are determined, for those of billions of years' duration obviously cannot be observed directly. From the weight of a given sample of the material and the known weight of one of its atoms (its chemically determined atomic weight), the number of atoms in the sample, a very large number, may be determined. By counting the number of alpha particles emitted in some convenient time, perhaps in an hour, the *rate* of radioactive decay may be determined. Taking into account the law of radioactivity, the time required for half the sample to decay may be calculated.

Of the great number of other advances wrought by quantum mechanics, some theoretical and some of a technological nature, the following are sufficiently important to merit brief discussion.

THE SOLID STATE

THE familiar materials in our environment are either solids, liquids or gases. Gases and liquids are, in general, much alike in their physical properties while solids show a tremendous diversity. One need but think of such familiar things as copper, glass, paper, hair, sticks and stones. The reason is that the molecules in gases and liquids are *disorganized*; they are free to move among themselves in a helter-skelter fashion. But the molecules of solids are bound together; they are *organized* in various kinds of groupings. While there are very few distinctly different kinds of disorganization, there are innumerable different patterns of organizations. The analysis of the solid state has been a major challenge to the ingenuity of scientists. It was only with the development of quantum mechanics that some of the problems posed by this complex state of matter came to be understood.

The *heat capacity* of certain solid substances had been inexplicable in terms of classical mechanics. The atoms in crystalline solids are held in their regular position by elastic forces and are constantly in a state of vibration about these positions, a vibration that becomes more extended at higher temperatures. The heat capacity, the heat energy needed to raise the temperature, is just the energy supplied to increase the vibration of the atoms. The amount of this energy required may be calculated by the laws of classical mechanics, and for many substances the experimentally measured values are found to be in close agreement with these calculations. There are, however, some puzzling exceptions, notably diamond, for which the measured value is far too small.

The explanation is that, as in the case of Planck's oscillators, the energy of the vibrating atoms is quantized; it can change only in discrete steps, steps which are greater for vibrations of high frequency. For many substances the frequencies are rather low, the steps are small, and the energy may change through many small increments nearly in the smooth manner of classical mechanics. In diamond, a crystalline form of pure carbon, the atoms are bound together by exceptionally strong forces, which accounts for diamond's great hardness, and they have a small mass, carbon being only number six in the periodic table of the elements. For these reasons the diamond atoms vibrate at an exceptionally high frequency, and a relatively large increase of energy is required to raise the vibration from one quantized state to another. When a diamond is heated, as by placing it in hot water, only a fraction of the atoms, those which happen to receive an exceptionally strong impact from one of the water molecules, take up any heat energy. This accounts for the abnormally low heat capacity. It might be said that, when a pound of diamonds is heated, only about three ounces of atoms actually get hotter.

Through the use of quantum mechanics metallurgy has been transformed from a craft to an exact technology. It has become possible to formulate new kinds of steel and their alloys (mixtures of several metals) that have greater strength, more resistance to corrosion and other desirable properties. Various improved magnetic materials have been developed, in particular new alloys for the construction of *superconducting electromagnets*, which do not require a constant supply of power to maintain the magnetizing electric current.

The greater understanding of solids has contributed to the development of the *laser*, a device which can produce light beams of enormous intensity capable of burning instantly through the most refractory materials. Of greater significance is the fact that a laser beam is *coherent*, that is, the whole group of waves in the beam are oscillating in unison rather than in the random fashion of ordinary light beams. This property gives laser light beams unique and valuable properties, for example, the ability to carry messages like radio waves, that are just beginning to be explored.

But the one new advance in the physics of solids which has had the most far-reaching consequences in science, in technology and industry, and indeed in society as a whole, is in the field of *semiconductors*, crystalline materials whose electrical conductivity lies between that of metals, which are excellent conductors, and substances such as glass and porcelain, which do not conduct the current at all. These are the materials used to produce *transistors* and other electronic devices which are essential for the construction of miniature electronic circuits. Without *solid-state* circuits the program of space exploration beyond the earth's atmosphere would be well-nigh impossible.

Of far greater import has been the development of *electronic computers*, whose rapid improvement in efficiency and sophistication has depended entirely on the use of solid-state electronic devices. Computers have become an indispensable research tool in physics, chemistry, biology, medicine, astronomy and all the other basic and applied sciences. They have made it possible to deal with problems that would otherwise be prohibitively difficult and time consuming.

Computers are taking over the operation of automated factories and chemical plants. They are accelerating and expanding the processes of design and development, management, control and decision making in manufacture, transportation and communication, education, government and finance. They bid fair to relieve human beings of all tedious and routine mental work, even as the invention of steam power and electric power has virtually eliminated back-breaking physical labor wherever they have become available. In the meantime the impact of computer technology has accelerated the pace of change in our economy, creating new jobs and destroying old ones, and raising a host of perplexing sociological problems.

SEEING WITH DE BROGLIE WAVES

SOME dramatic advances in experimental science, particularly in biology, have resulted from the invention of the *electron microscope*, whose operation depends directly on the de Broglie matter waves. Because of the limitations imposed by diffraction effects (page 45) it is impossible to obtain with the ordinary microscope, using visible light, a sharp image of details smaller than about one-ten-thousandth centimeter.

A tremendous increase of magnification became possible when it was discovered that the guiding matter waves of electrons can be focused to form a sharp image by electric and magnetic "lenses," electrically charged metal rings and coils of current-carrying wires, whose action is analogous to the glass lenses of microscopes. Since these matter waves can be produced readily with wavelengths a thousand times shorter than those of visible light, the attainable magnification is enhanced correspondingly. The magnifying power of the electron microscope exceeds that of the best light microscope as greatly as the latter exceeds the capability of the unaided eye.

Matter waves are not directly visible, but the electrons guided by them, impinging on a fluorescent screen, form a sharp visible image much as the picture on the screen of a television tube is produced. Minute details of biological structure, which could formerly only be conjectured, are clearly revealed on the electron microscope screen. Much new knowledge about important physiological processes has been obtained by such observations.

OUT OF THE ATOMIC SUBSTRATUM

In the familiar world of ordinary experience the description of mechanical phenomena given by the new mechanics is, as has been made evident repeatedly, not sensibly different from that given by the older classical laws. In broad areas of engineering and technology and in the conduct of many everyday affairs quantum mechanics can be safely ignored. There are, however, important and prominent happenings in the large world which are directly related to what is going on among the atoms; and to describe and understand these, quantum mechanics is indispensable.

All *chemical* phenomena are of this sort, for a chemical reaction occurs when atoms and molecules meet and interact. But the results may become conspicuously evident on a large scale. Chemical processes are not restricted to laboratories of chemistry or to the installations of chemical industries. They are proceeding unceasingly in the soil, the air and the waters of the earth and notably in all living things.

Geologists, especially those working in geophysics, are interested in the peculiar properties of atoms at the high temperatures and enormous pressures existing in the depths of the earth. For atomic phenomena are involved in generating the forces which in past ages have raised mountains and have produced the geological formations found in the rocky crust of the earth.

In modern astronomy the principles of quantum mechanics, particularly as they apply to the study of atomic nuclei, have proven of primary importance. It has been found that atomic phenomena occurring inside the stars, under conditions far different from those obtaining on our earth, determine in large measure the course of cosmic phenomena.

Of greatest import for human affairs are the recent advances made in the life sciences, which have resulted directly or indirectly from the development of quantum mechanics. These have been achieved partly through new experimental procedures, such

as electron microscopy and artificial radioactive tracer techniques, but most notably through new theoretical methods of
analysis. Much of what goes on in living creatures results from
the physical-chemical interactions of their complex and labile
molecules. Therefore, quantum-mechanical methods are essential
for theoretical studies in the new field of *molecular biology*, in
fields such as genetics, virus infection, nerve activity, nutrition,
growth and ageing.

Dr. Warren Weaver, who during the many years he guided
research at the Rockefeller Foundation had the opportunity to
observe the progress of science and its impact on human affairs,
is convinced that the future progress of the life sciences lies in
the molecular approach. Concerning medical developments, he
writes (in *Scientific Research* of July 1967):

> Now we're really going to find out about immunology on the
> molecular level; and when we do the problem of disease will
> become not a problem of cancer, of polio, of influenza, of the
> common cold. . . . It will become just a problem of disease—just
> one problem.

He believes that a most promising field of research lies in
molecular neurophysiology. He says:

> I am convinced that molecular biology has now reached a point
> where it can make, over the next five to twenty-five years, some
> extraordinary advances in the general field of molecular analysis
> and interpretation of neurophysiological problems. We are going
> to learn something about the mind, the brain and behavior. For
> the first time we are going to learn it with our feet on the ground.

By this he means that we are going to be able to study mental
phenomena in terms of general principles based on sound experimental evidence and effectual theories.

This completes the story, as given by quantum mechanics, of
the structure and the activities of the atoms, or more specifically
of the atomic electrons. The discussion which follows is concerned firstly with *nuclear* physics, with experiments and theoretical studies aimed at describing the internal constitution of
atomic nuclei, and then with the totally unexpected results of
these investigations, results which opened up a new area of
research, the physics of *elementary particles*.

9 ⋮

Elementary Particles

DURING the first three decades of the twentieth century the realm of the atoms was broached and conquered. After Louis de Broglie had proposed the novel idea that waves are involved in the motions of material objects, the work of Heisenberg, Schrödinger, Born and others had brought forth an essentially complete account of atomic structure and dynamics. The quantum-mechanical theory which these men developed gives a satisfactory description of all those phenomena which may be ascribed to the activity of the electrons surrounding the atomic nucleus. It is natural, therefore, that the tiny nucleus itself should then have become the object of scientific scrutiny.

TRANSMUTATION OF ELEMENTS

AN important question for which nuclear research had to seek an answer concerns the nature of the force which holds the nuclear particles together. Since nuclei contain an excess of positive charge, presumably in the form of protons, the disruptive repulsion between these particles of like charge must be contained by a stronger attractive force, which cannot be electrical. The gravitational force is quickly ruled out; it is inadequate by a factor of more than a trillion trillion trillion. The nuclear binding force must therefore be of a sort quite unknown in the large-scale world. While it is strong, it should have an extremely short range. In his atomic bombardment experiments Rutherford had found that an alpha particle may approach a nucleus to within about 10^{-12} centimeter, a ten-thousandth of the atomic diameter, without experiencing any force other than the electrical repulsion of the positive nuclear charge. Evidently, the nuclear binding force does not reach beyond the closely packed particles of the nucleus itself, which explains at once why it remained quite obscure until experimental techniques were developed to study nuclei closely.

Rutherford himself was one of the first to attempt such a study. He reasoned that, by shooting alpha particles at the nuclei of light atoms, of low atomic number and small nuclear charge, he might bring them close enough so that the nuclear force could come into play. Accordingly, in 1919 he devised an experiment in which high-speed alpha particles were directed into a vessel containing nitrogen gas. He then found that very-high-velocity penetrating particles emanated from the nitrogen atoms, particles which proved to be protons. This surprising appearance of hydrogen nuclei was at first laid to a possible slight admixture of hydrogen in the vessel. But no matter how carefully the nitrogen was purified, the observed number of these nuclei remained undiminished. Rutherford was forced to conclude that he had actually transformed one kind of atom into another, that he had realized the centuries-old dream of the alchemists, the *transmutation of elements*.

The transmutation which Rutherford had observed may be described concisely by the formula:

$$_7N^{14} + _2He^4 \rightarrow _8O^{17} + _1H^1$$

Here the letters stand, respectively, for nitrogen, helium, oxygen and hydrogen, the atoms whose nuclei are involved in the transmutation (alpha particles being helium nuclei). The formula indicates that the nuclei of nitrogen and helium, approaching to within the range of the nuclear force, fuse to form a compound nucleus which immediately ejects a proton at high velocity. This ejection (artificial radioactivity) of a particle by an unstable nucleus, formed through nuclear interaction, has already been described in the discussion (page 112) of the Gamow nuclear theory.

The subscripts on the letters indicate the nuclear charges, in electron units $(+e)$ of positive charge. The superscripts are the nuclear masses in *atomic mass units* (AMU), defined as one-sixteenth the relative mass of oxygen in the periodic table of elements. (The oxygen produced in this transmutation is not the most abundant isotope, of 16 AMU, but a rare one of mass 17.)

The sums of the subscripts and of the superscripts balance on the "before" and "after" sides $(7 + 2 = 8 + 1$ and $14 + 4 = 17 + 1)$. The agreement in the sums of the subscripts is precise because the electric charge values of the nuclei are precise multiples of the electron unit. However, the energy changes in nuclear interactions are large enough to produce appreciable equivalent changes of mass, and the superscripts are understood to represent only the nearest integer values. (If the relative mass of $_8O^{16}$ is taken to be exactly 16.00000, the actual values for $_1H^1$ and $_2He^4$ are, respectively, 1.00814 and 4.00388.) Here the sum of the mass on the "after" side is slightly less than on the "before" primarily because of the energy carried off by the high-velocity proton.

The unforeseen discovery of transmutation aroused wide interest, and it was soon followed by numerous other investigations of the effects of alpha-particle bombardments. Of particular interest is the experiment with beryllium conducted by the team of Jean Frédéric Joliot (1900-58) and his wife Irène (1897-1956), daughter of Pierre and Marie Curie, the pioneers in research on radioactivity. They detected a mysterious, highly penetrating radiation emanating from the beryllium target. Since it was found to have no electrical charge, they surmised that it might be a stream of very-high-energy photons. When, however, Sir James Chadwick, a young English experimental physicist working in Rutherford's laboratory, repeated the experiment, his careful measurement of the energy and momentum transfers involved when this radiation impinged on hydrogen nuclei led him to conclude that he was dealing with a new kind of elementary particle. This was found to have a mass nearly equal to that of the proton, and being without electric charge, it was called the *neutron*. As was explained previously (page 35), this particle has been found to be a constituent of all atomic nuclei (except that of hydrogen).

In the publication, in 1932, announcing the discovery of the neutron, Chadwick wrote: "It is of course possible to suppose that the neutron may be an elementary particle. This view has little to recommend it at present." This attitude is typical of the reluctance by many physicists at that time to admit a third kind of particle, besides the proton and electron, as a fundamental constituent of the universe. It was suggested that the neutron might be a collapsed hydrogen atom, a closely bound proton-electron pair. But further studies showed that this idea is untenable, that the neutron must be accepted as a distinct entity.

It would seem fantastically improbable that, although everyone is surrounded by tons of neutrons in his immediate environment, no one should have been aware of their existence before 1932. This is easily explained by considering what an atomic nucleus "looks" like to a freely wandering neutron. For a positively charged particle the nucleus is "a well surrounded by a mound" (page 110). The "mound," the repulsion which the positive nuclear charge exerts upon the approaching particle, normally keeps it from getting too close to the "well." But for the neutron the nucleus is "a well without a mound"; and it easily gets close enough to "fall into the well," to be pulled into the nucleus by the strong nuclear force. Thus neutrons have a slim chance of remaining at large. Only after it became possible to shoot at nuclei with extremely energetic bullets could neutrons be blasted out into the open and detected for a moment as free particles.

The early experiments in nuclear bombardment were done with alpha particles, the high-energy bullets provided by nature in radioactive atoms. Later, large and powerful machines, *particle accelerators,* were built to raise protons and other charged particles to energies even exceeding those of nature's bullets. Using these, it was found that each of the many existing kinds of nuclei could be transformed into other kinds, some of which had not been known before. In particular, a large number of new isotopes of elements were produced. For example, it was found that, in addition to the two well-known isotopes of chlorine of atomic weight 35 and 37 (page 36), there are no less than seven others ranging in weight from 32 to 40. The lightest of these nuclei has 15 neutrons, the heaviest 23. All of them contain 17 protons and collect 17 electrons to form atoms having the chemical properties of chlorine. The seven newly discovered isotopes are unstable and revert spontaneously to other stable kinds of atoms by radioactivity.

Two formerly unknown elements have been produced by nuclear bombardment: technetium (atomic number 43) and promethium (atomic number 61) having, respectively, 15 and 16 isotopes, all of them unstable. Furthermore, bombardment experiments have generated since 1940 a number of *transuranium* elements with atomic numbers higher than 92. Before these appeared, it had been assumed that, for reasons unknown, nature could not pack more than 92 protons into one nucleus. The first of these new elements (atomic number 93) was named *neptunium* after the planet Neptune, whose orbit lies beyond that of Uranus; and the next (atomic number 94) took its name from the planet Pluto. There proved to be many more transuranium elements than there are planets beyond Uranus so that this scheme could not be continued. Largely through the work of a group of scientists at the Radiation Laboratory of the University of California at Berkeley, under the direction of the physicist Ernest Lawrence (1901-58), the list was extended one by one to number 103, fittingly named *lawrencium*. Still higher numbers are theoretically possible, but have not yet been achieved. All of these elements are radioactive with short half-lives and are therefore not found on the earth today. Uranium is also radioactive, but its half-life is roughly equal to the estimated age of the earth. Therefore, about half the uranium present when the earth was formed is still here.

In the intensely hot interiors of the stars, the thermal agitation of the atoms is so violent that their nuclei are stripped bare of electrons and are driven, in spite of their mutual repulsion, into direct contact with one another. This results in the regrouping of protons and neutrons to create new kinds of atomic nuclei. From evidence obtained through astronomical observations and nuclear collision experiments performed in physics laboratories, it appears that the enormous outpourings of energy by the sun and all the other stars have their source in these nuclear processes, and that in this manner the heavier elements, helium, carbon, iron and all the rest are forged within the stars out of hydrogen, the primordial stuff of the universe.

Whenever lighter nuclei merge into heavier ones, energy is released equal in amount to the energy which would be required to separate these lighter parts again. Since the nuclear forces binding them together are very strong, the amounts of energy involved are large, sufficient to account for the energy dissipation of stars over periods of billions of years. Furthermore, because of this energy release the mass of the newly formed heavier nuclei is always slightly less than the sum of their parts (compare page 122).

Many of the problems in modern astronomical research, particularly in *astrophysics,* which deals with physical phenomena in the stars, could not have been attacked or even formulated but for the development of nuclear physics. Its importance for the affairs of the universe at large is emphasized (in the May 1967 issue of *Physics Today*) by Victor Weisskopf, head of the Massachusetts Institute of Technology Department of Physics, thus:

> However, there are places in the universe where excited nuclei occur in large numbers; the natural habitat of nuclear structure physics is the interior of stars, and there are many stars; there are probably more stars than planets. Man has in fact created here on earth a cosmic environment, and this is what we are investigating. To have nuclear reactions going on in the laboratory—the same reactions that go on in the middle of the sun and the middle of Sirius—is a far more impressive human achievement than those trips to the moon, now so much advertised.

PARTICLES AND MORE PARTICLES

THE year 1932, in which the neutron was discovered, marks an important milestone in the progress of physics. This discovery of a new elementary particle, but the first of many more to come, ushered in a new branch of physics, *particle* research or *high-energy* research (the two are in many ways synonymous), which has come to exceed in scope and importance the *nuclear* research from which it arose.

The experimental investigation of new elementary particles and of their many and complex interactions has occasioned an unprecedented commitment of scientific manpower and of technological and financial resources. The theoretical study of the many new experimental findings has occupied the majority of physicists working at the forefront of theoretical research.

All of this tremendous activity is but the latest development in the age-old quest after the nature of matter, the stuff of which the universe is constituted. The great apparent variety and complexity of material substance has always posed a challenge to natural philosophers. Attempts to meet this challenge have led to various formulations of the general thesis that the countless forms of matter are but the varied combinations of a lesser number of distinct basic constituents.

The natural philosophers of ancient Greece developed two such conceptions: the "atoms" of Democritus and the "four elements" (earth, water, air and fire) of Aristotle. Democritus was vague about the number of different kinds of atoms; but it seems that he conceived of this number as being quite large, perhaps in the hundreds. Because of Aristotle's great authority the scheme of the four elements dominated philosophic discussions throughout the Middle Ages and the Renaissance. With the beginnings of true scientific work, after 1600, the atomic hypothesis gained the ascendancy. The contributions of many eminent men in the succeeding three centuries led to the recognition of some ninety distinct kinds of atoms, arranged in an orderly manner in Mendeléeff's table of elements. This was widely accepted as the ultimate description of all material elements.

However, the work of Thomson and Rutherford produced a further remarkable simplification, the reduction of all matter to but two kinds of elementary entities, the electrons and protons. But soon the picture was to be complicated again. After 1932 more and more new kinds of elementary particles were discovered in rapid succession. By 1957 the number stood at thirty.

Because of the environmental conditions existing on earth, these newly discovered particles can be brought into being only under very special circumstances. Furthermore, since they are highly unstable and exist for less than a millionth of a second, they are exceedingly rare; their contribution to usual familiar happenings is negligible. In the universe at large, however, particularly in distant galaxies where phenomena of great violence are occurring, they are no doubt of more importance.

To most physicists, persuaded that nature should be basically simple, the concept of a universe comprising a large number of distinct and apparently unrelated entities is intolerable. Various attempts to find a lesser number of more elementary parts, of which all of the many particles might be constituted, have, however, yielded surprising results. Some experimental findings have been interpreted to imply that, however difficult it might be to visualize this intuitively, each of the many elementary particles is somehow constituted of all the others, no one of them being more elementary than the rest. There are many examples of a given particle "coming apart" into two or more others; but some of the "parts" are found to come apart again, and one of *these* parts may be the original given particle. In fact, many particles can come apart in various ways to yield different kinds of particles; but there is good evidence that the parts do not exist in the whole before it separates. A considerable body of experimental evidence is explained in terms of the idea that particles are, in general, in a rapid state of fluctuation between being just themselves and being themselves plus a variety of other particles.

All of these puzzling observations, which will be given due consideration later, serve to show that in the weird world of elementary particles many conceptions derived from familiar experience are no longer valid. The definitions of an elementary entity as one which cannot be separated into lesser parts, and of a compound entity as one which may be visualized intuitively as consisting of separable parts, seem to have no meaning for particles. For these reasons the term "elementary" has fallen into disuse, and these most recently recognized constituents of material substance are referred to simply as *particles*. (Note that here the connotations of this term are rather different from those of its previous uses, as in the discussion of the wave-particle dichotomy.)

DISCOVERING PARTICLES

THE "discovery" of the numerous particles in the two decades following the identification of the neutron is perhaps better described as creation. With few exceptions the mean lifetimes of these particles are so exceedingly short, less than a millionth of a second, that they can be detected only just after they have come into being. The process of creating particles depends on the relativity relation (page 73) between mass and energy ($E = mc^2$), which implies that mass can be created out of energy. To create a particle, it is necessary, however, to have the requisite energy in a highly concentrated form, either as kinetic energy of another particle moving at a very high velocity or as an extremely high-energy photon.

When a particle is speeded up by the action of a force, the work done by the force appears as increased kinetic energy of the particle, its mass being enhanced by an amount equivalent to this increase of energy. At all ordinary velocities the mass increase is inappreciable. Even at one-thousandth the velocity of light (186 miles per second), high by ordinary standards, the increase of mass is less than one part in a million, hardly a measurable amount. However, for a particle moving at 99 percent of the velocity of light, there is a *sevenfold* increase of mass. The velocity of light in free space (186,000 miles per second) is an ultimate velocity, which material objects may approach but never attain completely. As a particle is pushed to a velocity very close to this ultimate limit, the increase in mass becomes very great. In particle experiments enhancement of mass by factors of hundreds or even thousands is not unusual. Photons, of course, do move at the velocity of light; their mass is due entirely to their energy.

When speaking of the mass of particles, it is necessary to distinguish between their *rest mass,* usually designated as m_0, and their *moving* mass m, also called *relativistic* mass. In all of the following discussions the subscript zero will be omitted and all references to the mass m of particles will be understood to be to the rest mass unless specifically stated otherwise. The increase of mass as the velocity v approaches the velocity of light c is given by the formula:

$$m = m_0/\sqrt{1 - v^2/c^2}$$

When high-energy particles are slowed down, usually by making them collide with stationary ones, they give up some of the additional mass they had acquired. Under proper circumstances this mass may go into the creation of new particles. The net result is that energy has been used up and particles have been created. If, as is frequently the case, the newly created particles are unstable ones, they break up quickly, sometimes through several intermediate unstable stages, to final, stable forms.

The most convincing examples of the conversion of energy to

mass are the interactions of high-energy photons with other particles. Their "rest mass," if this term were applicable, must be taken as zero. In colliding with material particles, they may disappear altogether, giving up all of their energy to the creation of new particles.

Early experimenters made use of the high-energy particles provided by nature in the *cosmic rays*, showers of particles, mostly protons together with a few heavier nuclei which, after being endowed with tremendous energies by cosmic processes not yet fully understood, fly into the earth and collide with the nuclei of atoms high in the atmosphere. Here they give up most of their energy and are caught up in a variety of interactions resulting in the creation of numerous particles, some of which reach the surface of the earth as *secondary* cosmic rays. Primary cosmic rays may be detected by climbing high mountains or by sending balloons equipped with suitable scientific apparatus to high altitudes. In recent times improved studies have been made in artificial satellites circling the earth high above the atmosphere.

While a considerable number of important discoveries have resulted from cosmic-ray studies, the interpretation of observations is complicated because this natural source of concentrated

Figure 25. Electric field in the space between two plates bearing positive and negative electric charges.

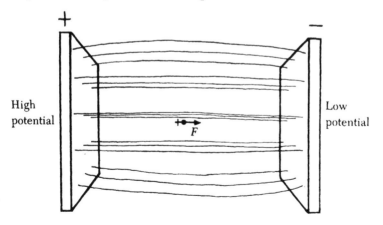

high energy is unpredictable as to time and place and, more importantly, because the energy varies greatly and is difficult to assess and impossible to control. Therefore, most of the solid progress in particle research has resulted from the use of *particle accelerators*, huge machines in which known kinds of particles may be given proper amounts of energy for specific purposes. The energies found in cosmic rays, however, exceed by far the utmost achieved by man-made machines.

PARTICLE ACCELERATORS

THE concepts involved in the action of particle accelerators follow from the basic fact that electrical charges of like sign repel while those of unlike sign attract each other. If two metal plates are set up, as shown in Figure 25, one charged positively and the other negatively, a positively charged particle placed in the space between them will experience an electric force *F* directed away from the positive and toward the negative plate. This situation is described by saying that an *electric field* exists in this space, the direction of the field being everywhere in the direction of the force on the positive charge. The electric field is indicated by *lines of force*, the fine lines in the figure, which start at positive and end at negative charges.

The positive plate and points near it are said to be at a *high electrical potential;* near the negative plate there is a region of *low electrical potential*. The *difference* of potential between the two plates is expressed in *volts*. One hundred volts is a moderately high difference of potential. The electric current, a stream of electrically charged particles, passing through a man's body under the influence of the electrical force due to this difference of potential may produce a dangerous electric shock.

A freely moving electrically charged particle, "falling" from a place of high to one of low potential, acquires kinetic energy, much like a stone falling from a higher to a lower point in the earth's gravitational field. The energy acquired by any particle carrying one electronic unit of charge while falling through a difference of potential of one volt is called one electron-volt (ev) of energy. After falling through a difference of potential of a million (10^6) volts, its energy is a million electron-volts (Mev); with a billion-volt (10^9) difference of potential it acquires a billion electron-volts (Bev) of energy. Electrons, and all other negatively charged particles, fall "uphill" from low to high potential, but the acquired energy is the same.

The energies of high-velocity particles are commonly expressed in Mev or Bev; and because in particle research energy and mass are being converted, one into the other, it has become customary to express the values of particle masses in these same energy units. The rest mass of the electron, for example, is given as 0.51 Mev. Cosmic-ray particles have been detected with energies as high as billions of Bev. When these particles collide with the nuclei of atoms in the atmosphere, they not only blast them apart into free protons and neutrons, but create showers of additional new particles out of their enormous store of energy.

All of the various machines designed to accelerate electrons, protons and other charged particles to high energies operate on the basic principles just discussed. They differ, however, in the ways in which the accelerating potentials are produced and made to act upon the particles. They also differ in size and cost, both having increased greatly as progressively higher energies have been attained.

The first successful accelerator was developed in 1929 at the Cavendish Laboratory by two hard-working young physicists, John Cockcroft and Ernest Walton, under the watchful eye and enthusiastic encouragement of Rutherford, who wanted high-speed protons to supplement alpha particles for his nuclear bombardment experiments. This machine obtained its high voltage by equipment similar to that used in X-ray machines (transformers and electronic rectifiers). It produced protons with energies up to 800,000 ev (0.8 Mev).

About this same time Robert Van de Graaff, working at the Massachusetts Institute of Technology, developed a machine reminiscent of the electrification devices used in the days of Benjamin Franklin. By dint of hard and persistent work he developed an exceedingly useful and reliable source of high voltage. Van de Graaff generators are capable of producing potentials up to about 15 Mev; but at these high voltages extreme precautions must be taken to prevent disruptive sparks, man-made bolts of lightning.

Further progress toward still higher particle energies was possible only through the ingenious device of accelerating particles, not by one tremendous push, but by a large number of lesser ones applied in succession. The first machine using this principle is the *cyclotron,* invented in 1932 by Ernest Lawrence at the University of California. Its operation depends on the action of *magnetic fields* on electric charges. Magnetic fields exist in the vicinity of *moving* electric charges, for example, near wires through which electric currents are flowing. Strong magnetic fiields are produced by *electromagnets,* coils of many turns of wire wound on iron cores to enhance the magnetic effects.

Whereas an electric field may produce a force on a particle in the direction of its motion and thus push it ahead and increase its energy, a magnetic field causes a force on a moving particle which is *sideways,* at right angles to the direction of motion. This force acts merely to deflect the particle into a circular path without changing its velocity and energy.

In the cyclotron the particles (usually protons) travel round and round through a magnetic field produced by a large and powerful electromagnet in an ever-widening spiral. Twice each time around they pass across a gap between two metal plates to which a difference of potential is applied momentarily at just the right instant to give a forward push in the direction of motion. In the first cyclotron protons made 500 passes over the gaps before being ejected, gaining 20,000 ev of energy at each pass for a total of 10 Mev.

The cyclotrons were superseded during the 1940's by bigger and more complex machines, the *synchro-cyclotrons,* capable of producing particle energies of hundreds of Mev. The billion-volt range was reached in the following decade by yet another type of accelerator, the *synchrotron,* in which particles travel round and round on a circular rather than on a spiral path. The first of these, a 1-Bev machine, was built in Birmingham, England, followed by one of 3 Bev at the Brookhaven National Laboratory on Long Island and one of 6 Bev at Berkeley. These were surpassed by a 10-Bev synchrotron put into operation in the Russian accelerator laboratory at Dubna in 1957. Two years later CERN (Conseil Européen pour la Recherche Nucleaire) activated a 30-Bev synchrotron at Geneva, Switzerland. A similar one, also of 30 Bev, began operation at Brookhaven in 1960. Russia took the lead in 1967 with the 70-Bev proton accelerator at Serpukhov; and in that same year plans were well under way for a European machine of 300 Bev and for one of 200 Bev in the United States so designed that its capacity might be doubled by later alterations. Even higher potentials, in the 1000-Bev range, are being considered.

The photograph, Figure 26, shows part of the large circular tunnel of the 30-Bev Brookhaven synchrotron which houses the "race track" through which the accelerated protons run. The track is actually a small tube, about half a mile in circuit, from which almost all the air is exhausted by vacuum pumps to prevent scattering and loss of energy through collisions with air molecules. In about one second the protons go around the track 300,000 times, guided by the magnetic fields of 240 strong electromagnets. Every time around they receive twelve pushes, each of about 8000 volts, giving them a total energy of 30 Bev. The

Figure 26. A section of the circular tunnel of the Brookhaven synchrotron.

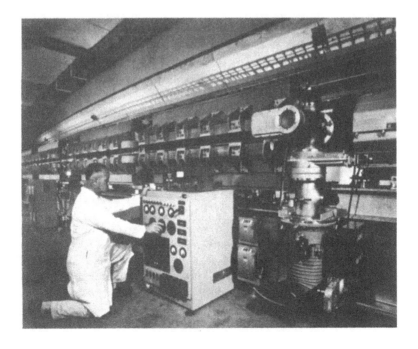

protons are sent into the accelerator in successive short pulses, each pulse being switched out of the track after reaching its maximum energy by a suitable deflecting magnetic field applied momentarily.

Electron accelerators operate in the same manner as those designed for protons. However, since electrons have a much smaller mass, they must be brought to higher velocities to attain the same kinetic energy. The maximum energy attainable by synchrotrons is less for electrons than for protons. According to a law of electrodynamics, charged particles traveling along a circular path radiate energy in the form of electromagnetic waves, and this radiation increases rapidly as the velocity of the particles is made greater. Therefore, one of the most powerful electron accelerators, at Cambridge, Massachusetts, produces electron beams with no more than 6 Bev of energy. Radiation losses are diminished if the track followed by the electrons is less sharply curved, that is, if it is a larger circle. For this reason a new electron synchrotron under construction at Cornell University will have a particle track one half mile in circuit.

Radiation losses may be avoided almost completely by building the accelerator with a long, straight track rather than a circular one. The principle of operation is somewhat different. The forward push on the electrons is provided by a rapidly advancing electric field. Its action has been likened to the forward thrust on a surfboard rider who skillfully maintains his board constantly on the sloping front face of the wave. The world's most powerful linear electron accelerator was placed in operation in 1966 at Stanford University. It is two miles long and accelerates electrons to an energy of 20 Bev.

Physicists working in the area of particle research have excellent reasons for pushing on to ever-greater particle energies. The history of particle physics, albeit brief, shows that each advance in available energy has yielded important new experimental findings. Repeatedly, the verification of tentative theoretical predictions has had to await the development of sufficiently powerful machines to put them to the test. For example, after it had been discovered that the electron has a double, the positron, it was postulated that a negatively charged proton might exist. But the experimental verification could not be attempted until the 6-Bev Berkeley synchrotron was in operation.

In the case of electrons higher energies provide a special experimental advantage. As has been stressed repeatedly in past discussions, it is impossible to distinguish details of objects that are smaller than the wavelengths of the waves used to observe them. One basic problem in particle research concerns the possible structure of the particles themselves. In attempting to study this structure, experimenters are trying to "see" objects about 10^{-13} centimeter across; and they would wish to distinguish details at least ten times smaller. The waves most suitable for this purpose are found to be the de Broglie matter waves of electrons. Recalling the de Broglie formula relating the wavelength λ of particles to their momentum $(\lambda = h/p)$, it is evident that greater momentum, and consequently greater kinetic energy, corresponds to shorter wavelengths. (At velocities approaching that of light, the increase of momentum is due to increase of mass rather than of velocity, but the de Broglie relation is still valid.) A wavelength of 10^{-13} centimeter corresponds to an energy of 1.2 Bev; a tenfold shorter wavelength requires a tenfold increase in energy. It is, of course, impossible to obtain an actual picture of a particle. But by bombarding protons, for example, by high-energy electrons and observing the details of the resultant scattering of the electrons, which is mediated by their matter waves, it is possible to make deductions about the shape and structure of the protons. Pioneer work of this kind was done by Robert Hofstadter at Stanford University with electrons of 200-Mev energy, and improved results will no doubt be obtained with the new 20-Bev accelerator.

The term "accelerator" for all of these high-energy machines is actually a misnomer, if *acceleration* is given its common meaning of an increase in *velocity*. In the Cambridge electron accelerator, for example, the electrons are injected into the track from a small linear accelerator at 0.99986 times the velocity of light and emerge at 0.999999996 times this ultimate velocity, an almost negligible increase. (In more easily comprehended terms, this increase in velocity is relatively the same as that between one automobile which can make a given trip in 2 hours and another faster one which completes the trip in 1 hour 59 minutes 59 seconds.) However, the increase in *mass* is considerable, from 60 times to 11,800 times the rest mass. This increase is the mass equivalent of the energy supplied to the electrons by the electrical forces driving them around the track.

STUDYING PARTICLES

To make observations on inconceivably tiny particles moving at enormous velocities and existing for less than a millionth of a second would seem utterly impossible. Yet all the known kinds of particles have been studied in considerable detail with regard to their mass, electric charge, spin and various other properties. This has been accomplished with a number of ingenious devices, including in recent years highly sophisticated electronic computers.

The simplest of these devices are the particle detectors or counters, which tell merely that a particle is present. They can locate a speeding particle within about one centimeter in space and to within a hundred-millionth second in time. The two principal types are the Geiger counter and the *scintillation* counter. The former consists of a small metal cylinder with an electrically insulated wire at its center to which a high voltage, nearly sufficient to cause an electric spark discharge, is applied. A single high-energy particle passing through the counter sets off a momentary pulse of electric current between the cylinder and the wire, which may be made to actuate a recording device. In the scintillation counter the passage of a particle through a special material produces a momentary flash of light, which is caught in a photomultiplier tube (page 62).

More information about the nature of particles may be obtained from devices which produce a trace of their *path*. Since particles are usually moving at velocities of about 10^{10} centimeters per second, they can traverse a path of a hundred centimeters even if they exist for no more than a hundred-millionth (10^{-8}) second.

One device of this kind is much like ordinary photographic film except that the emulsion is more heavily loaded with fine grains of silver bromide. A particle streaking through the film leaves a track of activated grains, which becomes a fine black line when the film is developed in the usual manner. Since films are light and compact and can make a permanent record of particles passing through them at unknown random moments, they are well adapted to the study of cosmic-ray phenomena in high-altitude balloons.

Longer tracks, which are immediately visible and may also be recorded photographically, are obtained in the *cloud chamber*. This device was invented in 1911 by the Scottish meteorologist C. T. R. Wilson (1868-1959). He was particularly interested in the formation of fogs and clouds which occur in moist air, when dust or smoke particles are present to act as centers about which the water molecules in the moist air may cluster to form fog droplets. He surmised that, in the very clean air at high altitudes where clouds form, the water droplets might collect around *ions*, air molecules which have become charged electrically by losing one of their electrons through naturally occurring high-energy collisions. When he tested this idea by passing alpha particles though air supersaturated with water vapor, he found that droplets do indeed form in thin visible threads of fog along the paths where speeding alpha particles have passed, leaving behind a trail of ionized air molecules. It occurred to Wilson that a small chamber full of cold moist air, on the verge of cloud formation, might be helpful in studying the motion of tiny invisible particles.

Cloud chambers have proven exceedingly useful tools in particle research, particularly when they are placed in strong magnetic fields created by large electromagnets. The deflection produced by the field, the curvature of the path, depends on the particle's momentum, a faster and more massive one being deflected less. If the velocity or the energy is determined from other observations, the mass may be deduced from the curvature of the path. The sign of the electric charge may be found as well since the magnetic field deflects positive and negative charges in opposite directions.

A similar device, which has several important advantages over the cloud chamber for the study of very-high-energy particles, was developed in 1952 by Donald Glaser, a young experimental physicist working at the University of Michigan. His *bubble chamber* works on a principle just opposite that of the cloud chamber; it depends on the change of liquid into vapor rather than the condensation of vapor into liquid. When a high-energy particle passes through a vessel full of liquid, which is at a temperature just slightly above the normal boiling point, it leaves behind a trail of closely spaced fine vapor bubbles forming a thin visible trace which persists long enough to be seen and photographed.

Since the molecules are far more closely spaced in a liquid than in gas, even very short paths are delineated in fine detail. Furthermore, a particle is slowed down much more effectively in passing through a liquid so that in a bubble chamber of practicable size, several feet across, the whole track may be observed to its termination. Liquid hydrogen is commonly used in bubble chambers because of its simple physical properties.

The more recently developed *spark chamber* contains two or more large metal plates to which a high voltage is applied an instant after a particle has streaked through. A bright spark, a miniature flash of lightning, occurs along the path where the particle has produced a trace of ionized air molecules. The high voltage is usually triggered by a scintillation counter which "sees" the particle just as it enters the chamber. Finally, there is the *streamer chamber,* which combines some of the advantages of both the bubble and the spark chambers. One of these highly complex particle detectors was built in 1967 and used to great advantage at the laboratory of the Stanford linear electron accelerator.

Ingenious combinations of components have been devised for various particular situations, as when a search is conducted for a relatively rare event which may be produced along with thousands of more common ones. To photograph the tracks of all of them, and then to inspect all of the pictures to find the rare ones, is extremely costly in time and labor. With the spark chamber, and also with the streamer chamber, it is possible to arrange a deflecting magnetic field and several counters, together with appropriate electronic devices, so that the chamber is triggered into momentary activity only when the one desired event occurs.

After this brief discussion of the experimental techniques developed in high-energy particle research, it is possible to consider in some detail the identification of the numerous new particles which occurred within the three decades following Chadwick's discovery of the neutron.

GREAT SCIENTISTS WHOSE

CONTRIBUTIONS LED

TO THE DEVELOPMENT

OF QUANTUM MECHANICS

Sir Isaac Newton (1642–1727). English scientist and mathematician.

More than any other single individual, Newton deserves to be called the founder of science in the modern sense. He made epochal advances in physics and mathematics and, even more important, he developed a philosophy that scientific pronouncements must be based not on rhetorical argument but rather on careful experimental observations, and that it is the proper function of science to describe rather than to explain natural phenomena. He set a course which, over a period of more than two centuries, produced a corpus of knowledge which vastly exceeded the scientific accomplishments of the human race in its entire previous history.

Thomas Young (1773–1829).   English physicist and physician.

In 1803, Thomas Young devised a simple experiment which showed clearly that light is propagated in a wave-like manner. He proved definitely and finally that light cannot consist of corpuscles moving in accord with the Newtonian laws of mechanics.

James Clerk Maxwell (1831–1879). Scottish theoretical physicist and mathematician.

The most important of Maxwell's contributions to physical science is his monumental electromagnetic theory of light in which he brought optics and electrodynamics together into a unified theoretical structure, rightly considered the crowning achievement of classical physics. He changed the concept of light from mechanical waves in a quasi-material ether to oscillating electric and magnetic fields. This development, based upon Michael Faraday's concept of electric and magnetic lines of force, is notable as the first important break from the materialistic-mechanistic views of Newtonian physics.

Sir Joseph J. Thomson (1856–1940). English experimental physicist.

Thomson obtained convincing experimental evidence that electricity is not a "subtle fluid" but consists of discrete particles, the electrons. He measured their mass and electric charge and demonstrated that they are universal components of all atoms. Thus he showed that the atoms are not themselves indestructible elementary components of matter and that they have an internal structure which might be amenable to experimental investigation.

Ernest Rutherford, First Baron Rutherford of Nelson (1871–1937).
British experimental physicist.

Rutherford was the first to succeed in exploring the internal structure of atoms. By experimental observations on the deflection of alpha particles passing through them he found that atoms consist of a massive positively charged nucleus surrounded by light, negatively charged electrons. He proposed the planetary atom model with electrons in orbits about the nucleus like planets about the sun.

Max Planck (1858–1947). German theoretical physicist.

Planck originated quantization, the salient concept of modern theoretical physics. He showed that the radiation of heat and light from hot bodies occurs in discrete packets or quanta of energy.

Albert Einstein (1879–1955). German-American theoretical physicist.

While Einstein is known primarily for his theory of relativity he also contributed importantly to other areas of physics, including light and atomic physics. He extended Planck's idea of quantization by showing that not only is light emitted and absorbed in quantized amounts but that it exists independently in the form of energy packets called photons.

Niels Bohr (1885–1962). Danish theoretical physicist.

Bohr richly deserves to be called the leader of atomic physicists. He combined the experimental findings of Rutherford with the theoretical concepts of Planck and Einstein to develop the Bohr atom model with its electrons, in discrete quantized orbits, distributed in successively larger shells about the nucleus, a model which accounted successfully for a vast number of experimental observations in chemistry and spectroscopy. Although later developments swept away the detailed picture of the Bohr atom model, the basic concept of energy absorption and emission through transitions between quantized energy states remained unchallenged.

Prince Louis de Broglie (*1892–1987*).   *French theoretical physicist.*

De Broglie is the founder of quantum mechanics. He extended the dual, wave-particle concept of light (resulting from the juxtaposition of Maxwell's electromagnetic waves and Einstein's photons) to material particles by endowing them with accompanying matter waves. His work eliminated serious logical inconsistences in atomic physics, and it altered the scientific concept of causality.

Erwin Schrödinger (1887–1961). Austrian theoretical physicist.

Using de Broglie's matter-wave concept, Schrödinger developed a logically consistent theory of atomic structure and dynamics. The quantized atomic energy levels, which had to be introduced *ad hoc* into the Bohr atom model, follow naturally from the basic concepts of Schrödinger's theory.

Werner Heisenberg (1901–1976). German theoretical physicist.

The results which Schrödinger derived from an intuitively visualized atom model, as a pattern of waves, Heisenberg obtained by rigorously excluding all mental constructs and building a theory based entirely on observable phenomena, on the measured values of the frequencies and intensities of the light emitted by the atoms. By probing further into problems of experimental observation and measurement in the realm of atoms, Heisenberg evolved his principle of uncertainty, which has had wide repercussions in science and philosophy.

10

Organization of Particles

PHYSICISTS have had a considerable measure of success in sorting out the many known particles in an orderly fashion. It has not been possible to arrange them all according to a single scheme, such as Mendeléeff achieved for the atoms. Whereas all atoms are built in a consistent manner of the same kinds of parts, particles are perplexingly diverse.

The first attempt at organization was in terms of a very obvious property, the *mass*. It seemed in fact that nature had arranged particles with respect to this property into three well-separated families, the *leptons, mesons and baryons,* names derived from Greek words meaning, respectively, "thin" or "small," "middle" and "heavy." All the known particles, with the exception of a most recently discovered group, which will be considered separately, are given in Table 3. Only four of these particles, the photon, electron, proton and neutron, have been discussed previously; the rest will be covered now.

The columns give the name and usual symbol, the spin s in Bohr angular momentum units (page 57), the rest mass m in Mev, the mean lifetime in seconds and three other properties still to be explained, as well as the antiparticles, also considered later. The superscript attached to each symbol is the conventional way of indicating the electrical charge, in electron charge units; where

Table 3. Table of elementary particles.

Family	Name	Symbol	Spin s	Rest mass, m	Mean lifetime	Strangeness, S	Isotopic spin, I	"Vertical" component, I_3	Antiparticle
	Graviton		2	0	∞				
	Photon	γ	1	0	∞				γ
Leptons	Electron	e^-	1/2	0.51	∞				$\overline{e^+}$
	Neutrino	ν_c	1/2	0	∞				$\overline{\nu_c}$
	Muon	μ^-	1/2	106	1.5×10^{-6}				$\overline{\mu^+}$
	Neutrino	ν_μ	1/2	0	∞				$\overline{\nu_\mu}$
Mesons	Pion	π^+	0	140	1.8×10^{-8}	0	1	1	π^-
		π°	0	135	0.7×10^{-16}	0	1	0	π°
		π^-	0	140	1.8×10^{-8}	0	1	-1	π^+
	Kaon	K^+	0	494	1.2×10^{-8}	1	1/2	1/2	$\overline{K^-}$
		K°	0	498	6×10^{-8} 1×10^{-10}	1	1/2	$-1/2$	$\overline{K^\circ}$
	Eta	η°	0	548	$\sim 10^{-16}$	0	0	0	η°
Baryons	Proton	p^+	1/2	938.2	∞	0	1/2	1/2	$\overline{p^-}$
	Neutron	n°	1/2	939.5	1013	0	1/2	$-1/2$	$\overline{n^\circ}$
	Lambda	Λ°	1/2	1115	2.5×10^{-10}	-1	0	0	$\overline{\Lambda^\circ}$
	Sigma	Σ^+	1/2	1190	0.8×10^{-10}	-1	1	1	$\overline{\Sigma^-}$
		Σ°	1/2	1192	$\sim 10^{-20}$	-1	1	0	$\overline{\Sigma^\circ}$
		Σ^-	1/2	1196	1.6×10^{-10}	-1	1	-1	$\overline{\Sigma^+}$
	Xi	Ξ°	1/2	1311	1×10^{-10}	-2	1/2	1/2	$\overline{\Xi^\circ}$
		Ξ^-	1/2	1319	1.3×10^{-10}	-2	1/2	$-1/2$	$\overline{\Xi^+}$
	Omega	Ω^-	3/2	1676	$\sim 10^{-10}$	-3	0	0	$\overline{\Omega^+}$

none is given the charge is zero. Those particles whose lifetimes are indicated by the mathematical symbol for infinity (∞) are stable, that is, there is no reason to doubt that they last forever. That the neutron is not among the stable particles, though its lifetime is enormously long compared to the rest, is a fact which merits special discussion. It may be surprising that photons are listed as stable particles since they cease to be whenever light falls upon matter and is absorbed. The criterion for stability is, however, that a particle be able to exist forever *if left to itself.* To the best of our knowledge the photons emitted by the stars may travel through empty space forever if they do not encounter an atom to absorb them. Alone of all particles the neutral kaon has two different lifetimes depending on the circumstances of its decay. The *graviton*, related to the gravitational force, is still hypothetical but of sufficient interest to be included.

A cursory inspection of the table reveals some obvious regularities. The spins have characteristic values for each of the three families (with the exception of the omega), and similar regularities occur among some of the other properties. But there are also some puzzling details. There is no apparent regularity in the sequence of mass values nor any easily discernible order in the distribution of charge values among the various kinds of particles.

THE LEPTONS

IN 1930 the English theoretical physicist Paul Dirac developed an important theory concerning the properties of electrons, in which he showed that there should be symmetry with respect to the charge, that there should be positive as well as negative electrons. Since the only positively charged particle known at the time was the proton, it was thought that this was Dirac's positive particle; and his theory was criticized for giving the "wrong" value of the proton mass. He was vindicated when Carl Anderson, working at the physics laboratory of the California Institute of Technology, obtained in 1932 the photograph, shown in Figure 27, of a particle track through a cloud chamber located in a magnetic field. The heavy bar across the chamber is a lead plate which the particle penetrated. The greater curvature of the path below the plate shows that the particle was moving more slowly there and, consequently, that it entered from above. The track is characteristic of an electron, but the direction in which it is deflected by the magnetic field shows that it was made by a positively

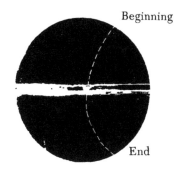

Figure 27. Photograph of the path of a positron in a cloud chamber.

charged particle. The full import of this discovery of the positive electron, or *positron*, was not appreciated until it was recognized as the first known *antiparticle*.

To predict the existence of a new particle before its discovery was a notable achievement. But remarkably, this was to be repeated twice over within the next few years. In 1931 Wolfgang Pauli had proposed an explanation for the puzzling observation that some of the beta rays emanating from radioactive nuclei have less than the expected amount of energy. He suggested that the missing energy might be carried off by elusive particles ejected along with the beta rays. This idea was not given serious consideration until the Italian physicist Enrico Fermi (1901-54) published, in 1934, a successful theory of beta-ray emission involving this new particle, which he called the *neutrino*. He predicted that it should have zero charge and zero rest mass, like the photon, but far greater penetrating power, so great that a stream of neutrinos might pass freely through the entire earth and through any apparatus set up to catch them. Neutrinos did indeed prove hard to catch. It was not until 1956 that two skillful experimenters, Fredrick Reines and Clyde Cowan, succeeded after three years of work in detecting neutrinos and verifying Fermi's prediction.

Like photons, neutrinos occur with a range of energies depending on the way they are produced. Those of greater energy have greater equivalent mass, greater momentum and shorter de Broglie waves. In fact, neutrinos and photons of the same energy have guiding waves of the same wavelength, light waves for the photons and matter waves for the neutrinos.

Neutrinos are produced in enormous numbers by processes occurring in the hot central core of the sun. Most of those streaming in upon the earth come from this source, more than a trillion (10^{12}) passing through every one of us in a second. Each neutrino carries energy of several million electron-volts; but since a person probably does not absorb more than one in an entire lifetime, they have no effect whatsoever. For neutrinos it is "daylight" around the clock; it might be said that for them the earth is like a sphere of clearest crystal.

About a tenth of all the energy radiated by the sun is in the form of neutrinos. But for some stars at certain stages of their life cycle neutrino radiation is probably the primary means of energy dissipation and the dominant influence in determining the course of their development. After being created deep within the center of a star, neutrinos penetrate quickly to the surface and start on their journey through space, a journey which may truly be called eternal, for the chance of their being absorbed by matter is less than one in a trillion trillion (10^{24}) over a period of ten billion years, the estimated age of the universe. It has been conjectured that nearly all of the neutrinos born since the dawn of creation are still coursing through space bearing most of the entire mass of the universe in the form of their energy.

Thus, neutrinos offer both the intriguing possibility of revealing some of the secrets of cosmology, of what has been going on over the ages deep within the interior of the stars, and the frustration of being almost impossible to catch. A start on "neutrino astronomy" was made in 1966 by setting up huge and elaborate detection equipment in a thousand-foot-deep gold mine in South Africa and in similar locations elsewhere, for neutrinos penetrate easily through a mile or two of solid rock which effectively stops all other interfering radiation.

Such experiments have the purpose of elucidating the role of neutrinos in cosmic phenomena. The inherent properties of these particles, and their mode of interaction with other kinds of particles, are being studied by experiments in which they are produced under controlled laboratory conditions.

A third prediction of an unknown particle was made in 1935 by a young Japanese physicist, Hideki Yukawa, while developing a quantum-mechanical theory of the strong force which binds the particles in atomic nuclei. On purely theoretical grounds he concluded that this particle should have a mass of about two hundred times that of the electron or one-ninth that of the proton. Because of the intermediate value of mass it was named the meson. About a year later a particle of this mass was actually found among secondary cosmic rays. Later, after a whole group of particles with such intermediate mass values had been discovered, it developed that the particle for which the name "meson" was coined does not belong in this group at all. Instead, it had to be placed with the leptons, for it was found to have all the properties of the electron except for its greater mass. Its name was changed to *muon* with the symbol μ (mu). Like the electron, the muon is associated with a neutrino which is designated by the symbol ν_μ to distinguish it from the electron neutrino ν_e.

Shortly after its discovery experimental evidence showed that the muon could not be the particle predicted by Yukawa. His theory implied that this particle should be affected strongly by the nuclear force while the muon could apparently collide freely with nuclei without being captured. For a number of years it seemed that Yukawa's theory was without experimental support.

The electron and muon, together with the two kinds of neutrinos, seem to complete the list of the leptons. None of the other kinds of particles identified in the thousands of experiments performed after the discovery of the muon could be assigned to this group.

THE MESONS

IT was fully twelve years before Yukawa's prediction was verified by the discovery of yet another particle having the predicted mass and also the required strong interaction with nuclei. For good measure it proved to be a triplet whose members have electric charges of $+e$, 0 and $-e$. These particles, called pi mesons or *pions*, are true members of the meson family. Pions, like muons, are commonly found among the products of cosmic-ray collisions at high altitudes. But unlike mesons, they cannot penetrate far downward through the atmosphere, so they were not discovered until detection equipment was sent to high altitudes.

The K-mesons, or *kaons*, were no doubt observed in cosmic-ray experiments by 1940; but they have such peculiar ways of changing into other kinds of particles that they were not fully understood until they had been studied intensively in the laboratory by particle accelerators. The eta meson, which exists only as a neutral particle, was unknown until 1960. Its extremely short lifetime, known to be about 10^{-16} second, sets it apart from the other mesons except for the neutral pion.

THE BARYONS

IN the first classification of particles the two earliest known heavy particles, the proton and neutron, were called *baryons* while the still heavier ones found in the 1950's were given the name of *hyperons*. Later, all were grouped together as baryons, and the term "hyperons" was no longer used. The omega baryon was not discovered until 1964. It is unique among the baryons in that it occurs only with negative charge and has an unusual value of spin (about which more will be said later).

ANTIPARTICLES

THE study of particles has brought forth the surprising realization that the stuff of the universe has a remarkable symmetry, that for all particles there are corresponding *antiparticles,* identical in every way except that their electrical charges and certain other characteristics are of opposite sign. Particles and antiparticles cannot exist together; their meeting results in mutual annihilation with production of particles of lesser mass.

The first observation of an interaction of this kind is the meeting of an electron and its antiparticle, the positron, with the production of two photons thus:

$$e^- + \overline{e^+} \to \gamma + \gamma$$

(The positron whose track is shown in the photograph obtained by Anderson, page 141, was not annihilated while traversing the cloud chamber because it was moving so fast that it did not remain near any one electron long enough to interact with it. After losing velocity by ionizing the atoms through which they pass, positrons do disappear promptly.)

The *bar* above the positron is the common designation of an antiparticle. The energy of the two photons is equivalent to the electron masses. Their wavelength is 2.4×10^{-10} centimeter, characteristic of very-high-energy X-rays (page 41).

The mutual annihilation of a proton and an antiproton may result in the creation of five pions:

$$p^+ + \overline{p^-} \to \pi^+ + \pi^+ + \pi^- + \pi^- + \pi^0$$

The pions, being unstable, decay spontaneously to other kinds of particles until only stable particles are left. The neutral pion changes directly to two photons:

$$\pi^0 \to \gamma + \gamma$$

The charged pions change to antimuons and muons plus neutrinos and antineutrinos:

$$\pi^+ \to \overline{\mu^+} + \nu_\mu \qquad \pi^- \to \mu^- + \overline{\nu_\mu}$$

The muons change finally to electrons, positrons, neutrinos and antineutrinos:

$$\mu^- \to e^- + \overline{\nu_e} + \nu_\mu \qquad \mu^+ \to \overline{e^+} + \nu_e + \overline{\nu_\mu}$$

All of these final products are stable particles.

All the antiparticles given in the table on page 140 have actually been observed, as have also most of the mutual annhiliation interactions. As seen in the table, most particles have distinct antiparticles. The photon is, however, identical with its antiparticle, as is also the neutral pion and the eta particle, while the two charged pions are the antiparticles of one another.

Antiparticles are in no way secondary to particles, for it is quite proper to think of particles as the antiparticles of antiparticles. The antiparticles of stable particles are also stable by themselves. The reason why the few antiparticles produced in physics laboratories cannot exist for long in an environment of particles is obvious. It is equally obvious why it would be exceedingly difficult to create whole antiatoms, consisting of antiprotons, antineutrons and antielectrons, in an apparatus made of atoms. A modest beginning was made in 1965 with the creation of an antideuteron, a nucleus consisting of a closely bound antiproton-antineutron pair.

Recent cosmological studies have led to the conjecture that the universe consists of matter and antimatter in equal amounts, these two kinds occurring as groups of galaxies and antigalaxies widely separated in space. It is highly improbable but not impossible that galaxies and antigalaxies might collide with catastrophic mutual annihilation and vast outpourings of photons and neutrinos. Astronomers have observed certain remote sources in space radiating tremendous amounts of energy which might be attributed to such a collision.

The discovery of antigalaxies could not be accomplished by means of the conventional astronomical observation of starlight nor by the newer techniques of radio astronomy, for antiphotons are identical with photons. The existence of antigalaxies might be established, however, through the detection of cosmic antineutrinos, because particle interactions are known which distinguish between neutrinos and antineutrinos.

Most of the neutrinos coming from our sun are produced by a complex sequence of nuclear interactions whose net result may be symbolized thus:

$$4_1H^1 \rightarrow {}_2He^4 + \overline{2\,e^+} + 2\,\nu_e$$

(Four atoms of hydrogen disappear and one helium atom plus two positrons and two neutrinos are created.) This same process occurs generally in stars consisting of the familiar kind of matter. In antistars an analogous process would produce antineutrinos. Astrophysicists have estimated that under favorable circumstances it should be just possible to detect neutrino radiation from distant regions of space. Thus, it appears to be only a matter of time before antigalaxies, if they exist, will be discovered.

The postulated dense swarm of neutrinos coursing through the universe (page 143) may well be half neutrinos and half anti-neutrinos. On the basis of theoretical studies it appears that mutual annihilation interactions between the two kinds (with production of photons) are so improbable that the loss, even in a time equal to the estimated age of the universe, is negligible.

There is one curious bit of evidence implying that a clash of matter and antimatter has actually occurred on the earth. In 1908 a spectacular explosion was seen in the sky above a remote region in central Siberia. A scientific expedition which reached the site many years later found trees blown down and scorched over a radius of thirty miles. The idea that a large meteor had struck the earth at that point was untenable because there were no remnants of a meteor to be found. Whatever it was that struck the earth had disappeared completely. No reasonable explanation was advanced until Willard Libby of the University of California, who had won the Nobel Prize in chemistry for his development of the radio-carbon method of dating archeological findings, made the ingenious suggestion that the explosion might have been caused by the impact of a piece of antimatter which had found its way to the earth from some remote region of space. Such an event would result in the production of a considerable amount of radio-active carbon, carbon atoms of atomic weight 14 with a half-life of 5760 years. These atoms, being chemically identical with normal carbon atoms of atomic weight 12, form carbon dioxide gas (CO_2) and, after being dispersed throughout the atmosphere, are incorporated into plants by the process of photosynthesis like normal CO_2. Libby and his associates studied the annual growth rings of a fir tree one by one, and found a slight excess of C-14 radioactivity in the ring of 1909 in comparison to those of the previous and the following years.

Be that as it may, it is a fact that, when matter is created out of energy in the laboratory, equal numbers of particles and anti-particles appear. It is not unreasonable to conjecture that this symmetry is characteristic as well for the universe as a whole.

11 $\overset{\circ}{\underset{\circ}{\overset{\circ}{\overset{\circ}{\circ}}}}$

Particle Characteristics

FORCES

In the familiar world of everyday experience *forces* are thought of as pushes and pulls. With particles the word has a broader connotation. When, for example, high-energy protons are directed at the stationary protons in a chamber filled with hydrogen, they may be merely scattered by the forces acting during collisions; or other phenomena, such as annihilation of the original particles and creation of new ones, may occur. In particle physics the word "force" indicates the cause of all these various interactions.

In general, when two particles meet and interact, new kinds of particles are found to leave the site of the interaction. It may happen, however, that the particles existing after the interaction are of the *same kind* as those which existed before. Since all particles of one kind are identical, the question as to whether they are the *same particles* is impossible to answer. It is more consistent to assume that annihilation followed by creation of new particles occurs in all interactions, with a certain probability that the new particles will be of the same kind as the former ones. In this case the interaction *could* be described as a scattering due to pushes or pulls. Thus, it appears that the familiar conception of force is but a special case of the more general annihilation-creation process found in the study of particles.

Classical physics recognized two kinds of forces: *gravitational* and *electromagnetic*. (Magnetic forces arise between electrical charges in motion and are therefore linked with electrical forces in a single term.) Particle research has revealed two additional kinds of forces called simply the *strong force* and the *weak force*, these noncommittal names being indicative of our ignorance concerning their nature. The former has already been encountered as the force which binds protons and neutrons in atomic nuclei. It is also the cause of numerous other kinds of particle interactions. The weak force gives rise to interactions of distinctly different kinds. Like the strong force, it has a range of action comparable to the size of nuclei. Therefore, neither of the two has any influence on the affairs of the large-scale world or even on extranuclear atomic phenomena. The latter are dominated by the electromagnetic force, which is also of importance in certain kinds of particle interactions.

The electromagnetic force is enormously strong relative to gravity, but large pieces of matter normally contain equal amounts of positive and negative charge so that all electromagnetic effects cancel out. The gravitational force acts as an attraction both for particles of like and unlike charge. Therefore, although the gravitational force is inherently very weak, the cooperative action of all the particles in a very large object such as the earth adds up to a strong total effect. There is no reason to doubt that the universal force of gravity acts between individual particles, but here its effect is negligible. If the strength of the electromagnetic force is given the value of unity, as a convenient standard of comparison, the relative strengths of the four known forces are related as follows:

Gravitational	10^{-36}
Weak	10^{-11}
Electromagnetic	1
Strong	100

The electromagnetic and the strong force stand together as far stronger than the other two. Compared to the electromagnetic, the weak force is less by a factor of a hundred billion; for the gravitational force the factor is an inconceivable trillion trillion trillion.

Since "force" has a complex meaning in particle physics, it is necessary to consider what is meant by the *strength* of a force.

This is taken to indicate how *quickly* it can bring about an inter-action; or more precisely, the strength of a force is a measure of the *probability* that it will produce an effect in a given brief interval of time. In this regard the strong force is indeed strong. Particles usually move with velocities comparable to that of light, and they can interact only while they are no more than about 10^{-13} centimeter apart. The time they have to interact is no more than one-hundred-billion-trillionth (10^{-23}) second. Yet strong force interactions do occur copiously in this incredibly short time.

The weak force acts in a more leisurely manner, requiring all of one-ten-billionth (10^{-10}) of a second to bring about an interaction. The relation between these two interaction times may be appreciated by noting that they stand in the ratio of a second to a million years.

Leaving gravity out of consideration, there are the following relations between forces and particle families. The leptons are not affected at all by the strong force, all of their interactions being brought about by the weak force. Both the mesons and the baryons respond to the strong force, so they have been grouped together under the common name of *hadons*. In addition, all charged particles are affected by the electromagnetic force. This force is also a factor in some of the interactions involving photons, which is not surprising since photons are created through the activity of electric charge.

The pions afford a simple example of the relation of forces to lifetimes of particles. The charged pions decay (are annihilated) by weak-force interactions to leptons, muons and neutrinos, with a lifetime of 10^{-8} second, while the neutral pion, which decays into two photons via the electromagnetic force, has the much shorter lifetime of 10^{-16} second.

The great difference in strength of the weak and the electro-magnetic forces accounts in part for the enormous difference in penetrating power of neutrinos and photons. In addition, there is another circumstance favoring the capture of photons, for this is brought about by the whole atom while the neutrino interacts only with the nucleus so that its "target" is smaller by a factor of about 10^{-10}. These two factors (10^{-11} and 10^{-10}) combine to a total ratio of 10^{-21} against neutrino capture.

An important example of an interaction due to the weak force is the spontaneous decay of a free neutron into a proton, an electron and an antineutrino:

$$n \rightarrow p^+ + e^- + \overline{\nu}_e$$

This decay process occurs in an average time of nearly twelve minutes, practically an eternity by standards of particle interaction. The time required for an interaction to proceed depends primarily on the kind of force involved but also on the amount of energy released through the disappearance of mass. Since the mass of the neutron exceeds that of the proton plus electron by less than one Mev, an exceptionally small amount for a particle interaction, the neutron is unique in the length of time required for its decay.

The exceptionally low driving energy of neutron decay accounts for the remarkable circumstance that neutrons, although they are inherently unstable, are nevertheless a constituent of stable matter. It might be said that the neutrons' very weak tendency to decay is inhibited by the strong force binding them in atomic nuclei. More correctly, the explanation is given by energy considerations. Within a stable nucleus the change of a neutron into a proton requires a certain supply of energy. The amount which would be made available by this change, the equivalent of the excess neutron mass over that of the proton plus electron, is less, *but only by a small fraction*, than the energy required to bring it about. If the mass difference were only slightly greater, the change could occur. All neutrons inside of nuclei would be converted to protons, or rather, no compound nuclei would have been formed, and the universe would have consisted of nothing but hydrogen. That the energy balance is a delicate one is evident from the fact that in the nuclei of radioactive atoms neutrons *do* change to protons with the subsequent emission of beta rays (page 109).

Such is the slender margin which has favored the creation of our richly varied environment and of ourselves to enjoy it.

CONSERVATION LAWS AND STABILITY

THE law of the *conservation of mass*, of great importance in classical science, asserts that mass is an inherent attribute of a given piece of material substance which remains constant in spite of all physical and chemical changes. A gram of water remains a gram when it is frozen to solid ice or changed to water vapor. Even if it is decomposed into its constituent chemical elements, the sum of the mass of the resulting oxygen and hydrogen is still one gram. In all the experimental observations of classical science no violation of this law was observed.

Of equal importance is the law of *conservation of energy*. This law states that, for all processes in which energy is changed from one of its many forms, such as mechanical, thermal or electrical, to another, the increase of one kind is exactly compensated by a decrease in one or more of the others. A simple example is given by the interconversion of the two kinds of energy, *kinetic* and *potential* energy, of mechanics. Kinetic energy is the energy possessed by a moving mass. An example of potential energy is the potentiality for doing work inherent in a mass which is in an elevated position in the earth's gravitational field. The weights of a grandfather's clock, by being wound up to the top of the case, are endowed with sufficient potential energy to run the clock for a week; and the water in an elevated lake has the stored-up potential energy to run a hydroelectric power station.

If a stone is tossed vertically upward, it leaves the hand with considerable kinetic energy; but as it rises, its velocity decreases until at its highest point it has lost all of its kinetic energy and is momentarily at rest. On the way up it gains potential energy, which exactly balances the loss of kinetic energy; on the way down the transformation goes the other way.

While the law of energy conservation places general limitations on the stone's motion, it does not determine the motion completely. If it were subject only to this law, it could, after leaving the hand, perform loops and other complicated motions just so it always "remembers" to slow down and speed up by the proper amount as it moves upward and downward. Obviously, additional laws (in this case Newton's laws of motion) are needed to assure predictable behavior.

This example shows a *general* characteristic of conservation laws. While they are of great importance in science, they determine only what may *not* happen, not what actually does happen in nature.

The laws of conservation both of mass and of energy were made obsolete when Einstein published the theory of special relativity. They were replaced by the one broader law of the *conservation of mass-energy*. It states that in the interconversion of mass and energy a change in one is exactly compensated by an opposite change in the other, the "rate of exchange" being 9×10^{20} ergs of energy for one gram of mass.

Because of the law of conservation of mass-energy all *spontaneous* transformations of a given kind of particle into others must proceed in the direction toward lesser mass, the loss in mass being compensated by the kinetic energy of the new particles. In the transformation, for example, of a neutron into a proton, electron and antineutrino (page 109), the three newly created particles have a total mass less than that of the proton. The reverse process, in which a proton is converted to a neutron, cannot proceed unless energy is *supplied* to make up the increase of mass. A stream of antineutrinos directed into a chamber containing hydrogen can supply the energy needed to produce the interaction:

$$p^+ + \bar{\nu}_e \rightarrow n + \overline{e^+}$$

The fact that protons and neutrons may be converted, one into the other, proves that a neutron cannot be thought of as a closely coupled proton and electron which "comes apart" into these two particles, for by the same token the proton "comes apart" into a neutron and positron, and so on *ad infinitum*. The only logical explanation is that each of these processes involves the annihilation of the original particles and the subsequent creation of the new ones.

Two other important conservation laws, valid in all of classical physics, have been found to hold as well, without exceptions, in all particle phenomena. These are the laws of conservation of *linear momentum* and of *angular momentum*.

Linear momentum p, usually called simply momentum, is already familiar as the product of mass m and velocity v. It is now necessary to add that momentum is a *directed quantity* (a *vector* quantity), that is, it is pointed like an arrow in the direction in which the mass is moving. Directed quantities have the important property that two of them equal in magnitude and opposite in direction cancel each other out (their sum is said to be zero), as is the case of two equal and opposite forces acting on the same object.

With this in mind it is possible to consider a simple case of conservation of momentum. A gun suspended on cords so as to fire horizontally is triggered by remote radio control so as to eliminate all outside forces. The law of conservation of momentum states that, under these conditions, the momentum of the gun-plus-bullet system will be conserved, will remain unchanged, when the gun is fired. Before firing, the total momentum of the gun and bullet, both at rest, is zero. If the mass of the gun is known to be ten times that of the bullet, it may be asserted with complete assurance that, when it is fired, it will recoil in a direction exactly opposite to the flight of the bullet and with a velocity exactly one-tenth of the bullet's velocity. The momentum of the gun (a tenfold mass multiplied by a one-tenth velocity) will then be exactly equal and opposite to that of the bullet so that the total momentum will be unchanged by the firing; it will still be zero. (Actually, the momentum of the ejected gas and smoke must also be taken into consideration.)

The principle of conservation of momentum was invoked to explain the curious observation that certain kinds of unstable atoms, produced by nuclear bombardment, are seen to take off suddenly, apparently all by themselves, leaving a visible track in a cloud chamber. It is quite as impossible for an atom to recoil without ejecting a particle as for a gun to recoil without firing a bullet. Since the ejected particle could not be detected, it was thought to be the hypothetical neutrino. Further studies showed that the nuclei of these atoms can capture one of their own electrons, thereby changing one of the protons to a neutron:

$$p^+ + e^- \rightarrow n + \nu_e$$

The neutron remains as a constituent of the new stable atom while the neutrino is ejected at high velocity. This experiment afforded indirect evidence for the existence of the neutrino before it was actually observed.

Because of conservation of momentum it is impossible for an unstable particle to decay spontaneously into *one* new particle. Since the transition would have to be to one of lesser mass, the new particle would take off with kinetic energy equivalent to the reduction in mass. While this would conserve energy, momentum would appear out of nowhere, which is impossible.

When a spontaneous decay process results in the creation of two particles, they fly apart in exactly opposite directions and with velocities inversely proportional to their masses. When more than two particles are created, the situation is more complex, but momentum is rigorously conserved in every case. If one of the new particles is uncharged and leaves no trace in the cloud or bubble chamber (since it cannot exert a force on atomic electrons to ionize atoms), its existence and its mass may be inferred from the principle of conservation of momentum.

Conservation of momentum explains why the mutual annihilation of an electron and a positron results in the production of two photons and not one. The two particles heading toward each other have zero total momentum and so have the two like photons leaving in opposite directions. A single photon of double the frequency would conserve energy but not momentum.

Further proof of the neutrino's existence was deduced from the law of conservation of angular momentum, which is related to rotation as linear momentum is to motion along a straight line. Angular momentum is also a directed quantity, its direction being given by the *right-hand* rule, as shown in Figure 28. If the curled fingers of the right hand indicate the direction of rotation, the

Figure 28. The relation between direction of rotation and direction of the angular momentum or spin arrow.

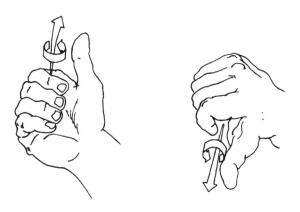

thumb shows the direction in which the arrow of angular momentum points. Arrows pointing in opposite directions indicate rotations going around in opposite ways.

As discussed before (page 60), the electron was found to have an angular momentum or spin. This is true as well of all the leptons and baryons; they all act as if they were spinning about an axis through their center. All of these spins are quantized, restricted to particular numerical values. Quantization of angular momentum was first proposed by Bohr (page 57) as the quantization condition which determines the successive discrete energy levels of the hydrogen atom. He postulated that the *orbital* angular momentum of the electron in its rotation around the nucleus must have a discrete sequence of values in units of $h/2\pi$, the *Bohr unit* of angular momentum. Later, it was found that the electron's *intrinsic* angular momentum about its own axis has the value $h/4\pi$, just *half* a Bohr unit. When this was found to be true for many other particles, their spin came to be designated simply as "one-half," the Bohr unit being understood. The spin of a particle has been found to be one of its important inherent properties, of more fundamental significance than its charge, as is evident in the organization of particles in the table on page 140.

The spin of particles is also *space-quantized* with respect to its orientation. This means that, when several particles are associated in an interaction, their spins are either in the same direction (parallel) or in exactly opposite directions (antiparallel). These two directions are usually designated by giving the value of the spin as $+\frac{1}{2}$ and $-\frac{1}{2}$.

The law of conservation of angular momentum or spin shows that a third particle, in addition to the proton and electron, must be involved in the neutron decay. By writing the spins below the particle symbols:

$$n \rightarrow p^+ + e^- + \overline{\nu}_e$$
$$\tfrac{1}{2} \quad \tfrac{1}{2} \quad -\tfrac{1}{2} \quad \tfrac{1}{2}$$

it is evident that spin is conserved, that the total spin is $\frac{1}{2}$ both before and after the decay has occurred. Without the third particle, the neutrino, the law of spin conservation would be violated. Since such an event has never been observed, physicists have complete confidence in this conservation law. They would be as astounded to discover a violation of it in particle interactions as they would to find some day that the earth is spinning on its axis the other way about, with the sun rising in the west and setting in the east. Thus, the observation of neutron decay was accepted as further evidence for the existence of the neutrino.

Particles obey in all of their interactions yet another classical conservation law, the law of conservation of *electrical charge*. Whereas this law was in classical physics an absolute prohibition of the creation or destruction of electric charge, it has a broader connotation in the physics of particles. It is possible to create a charged particle out of energy, provided that one of equal and opposite charge is created at the same time. Inspection of all the particle interaction formulas given thus far will show that charge is conserved, in this sense, in every case.

In addition to all of the above, particle interactions are restricted by additional conservation laws unknown to classical science. Two of these may be stated very simply. The law of conservation of *baryon number* asserts that the total number of baryons in the universe never changes. In the interpretation of this law each baryon is arbitrarily assigned a baryon number B of $+1$ while that of an antibaryon is given as -1; and the law implies that, whenever a baryon is created, an antibaryon must be created at the same time. The first observed antiprotons were created by bombarding hydrogen nuclei with high-energy protons:

$$p^+ + p^+ \rightarrow p^+ + p^+ + (p^+ + \overline{p^-})$$

The bombarding proton and the target proton survive and a proton-antiproton pair is created out of energy, with conservation of baryon number. This interaction, first observed in 1955, was followed a year later by the creation of the antineutron:

$$p^+ + p^+ \rightarrow p^+ + p^+ + (n + \bar{n})$$

In such proton collisions one or both of the original protons may disappear and be replaced by heavier baryons created out of the rest mass of the annihilated protons plus the additional energy of the bombarding proton, again with conservation of baryon number. For example:

$$p^+ + p^+ \rightarrow p^+ + \Lambda^0 + K^+$$
$$p^+ + p^+ \rightarrow \Lambda^0 + n + K^+ + \pi^+$$

(The reason why the kaons appear will become apparent in the discussion of "strangeness.")

Lepton interactions are restricted by an even more specific conservation law. Here again leptons and antileptons are given, respectively, the *lepton numbers* +1 and −1; and the rule is that the number of electron leptons and muon leptons must be conserved separately. This will be found true of all previously given interactions involving electrons and their neutrinos. In the decay of the muon to an electron:

$$\mu^- \to e^- + \nu_\mu + \overline{\nu}_e$$

the muon number is conserved by the appearance of the muon neutrino, and the creation of the electron is compensated by the electron antineutrino.

There is no conservation law for the number of photons and mesons. As seen above (and on page 145), pions and kaons may appear anew if there is sufficient energy available to create them. Photons appear and disappear freely in atomic emission and absorption processes and may disappear in the creation of mesons; for example:

$$p^+ + \gamma \to p^+ + \pi^0$$

Although particle interactions are restricted by the various conservation laws, the number of possibilities is still very great. Even the spontaneous decay processes starting with a single particle are more numerous than the number of different kinds of particles, for most of them may decay in more ways than one. The positive kaon, for example, can decay to all of the following products:

$$K^+ \to \overline{\mu^+} + \nu_\mu$$
$$K^+ \to \pi^+ + \pi^0$$
$$K^+ \to \pi^+ + \pi^+ + \pi^-$$
$$K^+ \to \pi^0 + \overline{e^+} + \nu_e$$
$$K^+ \to \pi^0 + \overline{\mu^+} + \nu_\mu$$
$$K^+ \to \pi^0 + \pi^0 + \pi^+$$

These are arranged in order of decreasing probability, and several more of lesser probability could be added.

Conservation laws have one consequence of primary importance. It is these laws which alone account for the very few *stable* particles and thus for the existence of the physical universe. The particles of *zero* rest mass, the photons and the two kinds of neutrinos, are stable because spontaneous decay processes must always proceed in the direction of lesser rest mass, and these obviously can proceed no further. In addition, the laws of conservation of electron and muon leptons prevent them from changing into each other. The electron is stabilized by charge conservation. Without this restraint it could decay into an electron neutrino plus a pair of photons. The proton is prevented from changing to a particle of lesser mass by the law of baryon conservation since it has the least mass of the particles in this family.

This completes the short list of stable particles except for the neutrons, whose peculiar way of achieving a spurious stability, by hiding in atomic nuclei, has already been discussed. All the other particles, the one heavier lepton, all the heavier baryons and all of the mesons, are unstable in spite of the many conservation laws limiting their decay processes.

CONSERVATION LAWS AND PROBABILITIES

Following the discussion of conservation laws in particle physics, it is instructive to consider anew the similarities and contrasts between the classical laws of the large-scale world and the laws of quantum mechanics, which were developed to deal with phenomena at the submicroscopic level. Classical physics recognizes two different categories of laws: the conservation laws, which place only general restrictions upon the nature of possible phenomena, and deterministic laws, which govern the course of phenomena in detail.

As soon as it became possible to make observations on the individual behavior of entities in the atomic realm, it became evident that these are subject only to laws of probability and that their individual behavior is unpredictable. It is important to realize, however, that conservation laws, which have limited predictive value even in the large world, are still valid for *individual* events in the world of submicroscopic things. For example, when an electron is poised at a high energy level in an atom, quantum mechanics can determine only the relative probabilities of its transitions to the various lower levels. But the experimental evidence shows that in any one transition energy is strictly conserved, the energy of each emitted photon being equal to the energy loss of the atom. If energy were conserved only on a statistical average, the emitted photons would not be limited to certain discrete values.

This is true generally of all conservation laws discussed thus far; they apply with equal force in the atomic, nuclear and particle realms as they do in classical physics. This is in harmony with the spirit of quantum mechanics, for it still leaves individual events unpredictable. In the various ways, for example, in which the positive kaon may decay to other particles (page 159), all the pertinent conservation laws are obeyed in each individual interaction. Yet it is impossible to predict which of these any individual kaon will follow. That there are statistical laws in force is evident; the first type of decay, for example, occurs in 59 percent of all possible ones.

In the large-scale world, conservation laws permit an *infinite* variety of possible happenings within their broad scope. Therefore, to have an orderly universe, it is essential that there be more specifically restrictive laws, such as Newton's laws of mechanics. With particles the conservation laws permit only a *finite* number of different events in a given situation. In fact, for most situations the combined restriction of several relevant laws limits the outcome to a few choices or even a single choice. Thus, it is possible to have a considerable measure of order in particle affairs without additional restrictive laws. Within the range permitted by conservation laws the rule is: whatever *can* happen *will* happen.

STRANGENESS AND ISOTOPIC SPIN

WHEN, during the 1950's, the kaons and the lambda, sigma and xi baryons were being discovered and studied, it became apparent that all of them exhibit a common strange behavior. In the light of previous limited experimental evidence, it had been assumed that a given particle is subject in *all* of its interactions to one and the same kind of force and that all of them should therefore occur

in about the same brief interval of time. Yet with these particles
the decay was found to proceed far more slowly than the interaction in which they are created.

Two examples of kaon and baryon creation have already been
given (page 158). There are numerous others, of which the following are examples:

$$p^+ + p^+ \rightarrow p^+ + \Xi^0 + K^+ + K^0$$
$$p^+ + \pi^+ \rightarrow \Sigma^+ + K^+$$
$$p^+ + \Lambda^0 \rightarrow p^+ + n + \overline{K^0}$$
$$p^+ + \gamma \rightarrow \Lambda^0 + K^+$$

Because of the manner in which these interactions are brought
about, it is certain that they occur in the very brief time characteristic of the strong force. Yet when these newly created particles
decay, they do so in the much more leisurely fashion of the weak
force. Decays such as the following:

$$K^0 \rightarrow \pi^+ + \pi^- + \pi^0$$
$$\Lambda^0 \rightarrow p^+ + \pi^-$$
$$\Sigma^+ \rightarrow n + \pi^+$$
$$\Xi^- \rightarrow \Lambda^0 + \pi^-$$

which involve only particles subject to the strong force and ought
therefore to occur in about 10^{-23} second, take instead 10^{-10} to 10^{-8}
second, trillions of times longer. Because of this these mesons and
baryons were given the common name of *strange particles.*

Strangeness was soon found to be more than a fanciful name;
it proved to be of real physical significance. Each of the strange
particles could be given a *strangeness number* (S), which, like
charge and spin, indicates a significant particle property. The
numerical value of S is related to the average charge of a *charge
multiplet*, a group of particles such as the two nucleons (p and n)
or the three pions, which differ only in electrical charge. The
average charge is the sum of the individual charges divided by the
number of particles in the multiplet. For the nucleons it is
$(1 + 0)/2 = 1/2$; for the pions it is $(1 + 0 - 1)/3 = 0$; and so on.
Twice the average charge is called the *hypercharge* (Y); and S
is related to Y by the formula:

$$S = Y - B$$

where B, the baryon number, is 1 for baryons and zero for mesons. (None of these properties have any meaning for the leptons.) As is true for Y and B, the value of S for antiparticles is the negative of its value for particles. For the nucleons and pions S is zero; for the other baryons, lambda, sigma, xi and omega, in order of increasing mass, the values are, respectively, -1, -1, -2, -3. The kaon is the only particle for which S has a *positive* value of $+1$.

The significance of strangeness for particle interactions lies in that it is subject to a peculiar conservation law. Conservation of strangeness holds in all *strong* force interactions but is violated by the *weak* force. In the creation of the strange particles by the strong force, strangeness must therefore be conserved; and this can happen, in general, only through *associated production*, in which the negative strangeness of the baryons is compensated by the positive value of kaons created in the same interaction. In the creation of the xi baryon (page 162), for example, by the collision of two protons, its value of $S = -2$ is balanced by the value $S = +1$ of the two kaons. As inspection will show, total strangeness is conserved in all the given examples of strange-particle creation.

After being created, the new particles separate at high velocity and must therefore decay individually. Leaving out of consideration the omega baryon, which was not discovered until 1964 and was unknown when the concept of strangeness was being developed, none of the strange baryons, with one exception, can decay spontaneously without violating the conservation of strangeness, for this annihilation does not set free sufficient energy to create antikaons. They can decay only via the weak force, which accounts for their remarkably long lifetimes. If the decay of the neutral lambda (page 162) to a proton, for example, could produce an antikaon in place of the negative pion, it could proceed far more speedily.

The one exception is the decay of the neutral sigma, with conservation of strangeness:

$$\Sigma^0 \to \Lambda^0 + \gamma$$

This is brought about by the electromagnetic force, which also obeys strangeness conservation, in about 10^{-20} second. (The rapid decay of the charged sigmas is blocked by charge conservation.)

Many processes of particle creation, which must be brought about in the exceedingly brief time in which only the strong force can act, are forbidden by conservation of strangeness. For example, an interaction such as:

$$p^+ + \pi^- \rightarrow \Lambda^0 + \pi^0$$

is in accord with the conservation of energy, momentum, spin, charge and baryon number; but it violates conservation of strangeness and is, in fact, never observed.

Particles have yet another property whose conservation law is related to forces in a complex manner. It is called *isotopic spin,* a curious name for a property which has no physical relation either to isotopes or to spin. The name was coined by Werner Heisenberg, who saw in the relation of protons and neutrons a loose analogy with isotopes, which differ in the number of protons and neutrons in their nuclei. The analogy to spin is an abstract, mathematical one. Just as the real spin of a particle may assume only certain discrete positions in space (up for spin $+1/2$, down for $-1/2$), so isotopic spin is limited to certain formal "orientations" in an abstract, nonphysical "space" of its own.

The values of isotopic spin (I) for the baryons and mesons are given in the table on page 140. (Isotopic spin, like strangeness, is of no significance for the leptons.) The "vertical" component of I, also called the "third component of isotopic spin" (I_3), is determined by the orientation of the isotopic spin "arrow." For $I = 0$, there is only a single position corresponding to $I_3 = 0$. With $I = 1/2$, I_3 may have two values, $+1/2$ and $-1/2$, for "up" and "down," while $I = 1$ permits three values of I_3, $+1$, 0 and -1, corresponding to the orientations "up," "horizontal" and "down." The value of I is related to the multiplicity (M) of a charge multiplet; it has the values 0, 1/2 and 1, respectively, for charge singlets, doublets and triplets. The values of I_3 are related to the electric charge of the individual particles in a multiplet; thus for the three pions, I_3 has the values $+1$, 0 and -1. As is true of electric charge, the values of I_3 for particles and antiparticles are opposite in sign.

The particle property I_3, like strangeness, is conserved in interactions produced by both the strong and the electromagnetic force. But conservation of isotopic spin itself is unique in being violated by the weak *and* the electromagnetic force. This peculiar circumstance is related to the fact that the strong force is indifferent to the values of charge in a multiplet (it acts in the same manner, between protons and neutrons), while the electromagnetic force does distinguish between the several multiplet members of different charge.

The concept of isotopic spin was developed while studying complex theoretical relations among particles and has proven to be of great value in understanding such relations. The numerical values of I and I_3 are related to baryon number B, strangeness S, particle charge Q and multiplicity M by the relations:

$$I = (M - 1)/2$$
$$Q = e(I_3 + B/2 + S/2)$$

RESONANCES

ANY system which can execute vibrations at a certain frequency responds strongly to a stimulus of this same frequency. A crystal goblet which "rings" with a certain tone when struck may be set into strong vibration by singing this tone into it. For tones of any other frequency or pitch the response will be less. This selective response to one particular tone is called *resonance*.

The concept of resonance has been extended to other phenomena in which there is an especially strong response to a stimulus of a particular value. Enrico Fermi, working at the University of Chicago in 1952, observed such selective effect when he bombarded protons in a vessel of liquid hydrogen with high-energy pions. He found that there is an exceptionally strong interaction between the protons and the pions when their energy is just 160 Mev. The importance of this observation was not appreciated until some years later, after many similar examples had been observed. It was then realized that Fermi had created a new and different sort of particle, the first of many more to come. All of them are created copiously when the total available mass-energy, the sum of the masses of the interacting particles plus the additional kinetic energy of the one used for bombardment, is just equal to the mass of the newly created one. Because they all have exceedingly short lifetimes, of about 10^{-23} seconds, there has been

some doubt as to whether they deserve the status of actual particles. Therefore, they have been called resonant particles or
resonances.

The resonance discovered by Fermi, designated by delta (Δ),
may be created by bombarding protons either with positive or
negative pions:

$$p^+ + \pi^+ \to \Delta^{++}$$
$$p^+ + \pi^- \to \Delta^0$$

The discovery of two additional deltas (Δ^+ and Δ^-) completed
the first and still the only known charge *quadruplet.*

The delta resonance may be created individually out of particles of zero strangeness because its strangeness is also zero.
Likewise, it can decay into protons and pions without violating
strangeness conservation, and it does so with the alacrity characteristic of strong force interactions:

$$\Delta^{++} \to p^+ + \pi^+$$
$$\Delta^0 \to p^+ + \pi^-$$

The relations showing the creation and the decay of the delta
resonances seem to indicate that nothing has happened but a
simple collision in which pions strike protons and rebound. This
does not explain why the interaction is particularly strong just
when the kinetic energy of the pions is 160 Mev. There are,
moreover, many examples of resonances in which the decay products are not the same particles as those which entered the
interaction.

It is possible, for example, to create a lambda resonance, with
a greater mass (1520 Mev) than the Lambda baryon, by bombarding protons with anti-kaons:

$$p^+ + \overline{K^-} \to \Lambda^{0*}$$

(The asterisk is commonly used to indicate a resonance.)

This resonance can decay in many ways, including the following:

$$\Lambda^{0*} \to p^+ + \overline{K^-}$$
$$\Lambda^{0*} \to n + \overline{K^0}$$
$$\Lambda^{0*} \to \Sigma^+ + \pi^-$$
$$\Lambda^{0*} \to \Lambda^0 + \pi^+ + \pi^-$$

Since the strangeness of the lambda resonance and of the lambda and sigma baryons and antikaons is −1, strangeness is conserved in all of these decays; and they are all brought about by the strong force in 10^{-23} second. None of these decay products, except the proton, are stable, however; and they decay further into protons, electrons, photons and neutrinos.

All of the resonances known in 1966, including meson resonances, are given in Table 4. It shows the average mass (in Mev) and the various electric charges for each multiplet, and the value of the spin s (in the usual Bohr unit), the strangeness S and the isotopic spin I. These three properties have proven useful in the classification of particles according to various theoretical schemes of organization. The table shows that a resonance, with the same symbol as that of a meson or baryon in the table on page 140, has like values of S and I but greater values of spin and mass. The symbol N corresponds to the nucleons, protons and neutrons. The s, S, I values of the delta resonances are not found in any of the baryons.

Table 4. Table of resonance particles.

Symbol	Rest mass	Charges	Spin, s	Strangeness, S	Isotopic spin, I
π	750	1, 0, −1	1	0	1
η	782	0	1	0	0
K	888	1, 0	1	1	1/2
η	1020	0	1	0	0
η	1250	0	2	0	0
Δ	1238	2, 1, 0, −1	3/2	0	3/2
Σ	1385	1, 0, −1	3/2	−1	1
Λ	1405	0	3/2	−1	0
N	1512	1, 0	3/2	0	1/2
Λ	1520	0	3/2	−1	0
Ξ	1530	0, −1	3/2	−2	1/2
Σ	1660	1, 0, −1	3/2	−1	1
N	1688	1, 0	5/2	0	1/2
Λ	1815	0	5/2	−1	0
Δ	1920	2, 1, 0, −1	7/2	0	3/2
N	2190	1, 0	7/2	0	1/2
Δ	2360	2, 1, 0, −1	9/2	0	3/2
N	2650	1, 0	9/2	0	1/2
Δ	2825	2, 1, 0, −1	11/2	0	3/2

Although the resonances are usually treated as a distinct category, it has become clear that they do not differ in any fundamental manner from the earlier known mesons and baryons. Their short lifetime is the logical consequence of the fact that there are particles of lesser mass and equal strangeness to which they may decay without violating any of the conservation laws governing strong force interactions. To recapitulate, these are the conservation laws of:

Mass-energy	Baryon number
Electric charge	Strangeness
Linear momentum	Isotopic spin
Angular momentum or spin	

The restrictions imposed by these laws limit sharply the number of possible resonance interactions. Yet their number is still sufficiently great to pose perplexing problems for theoretical physicists attempting to analyze them.

An experimental difficulty in dealing with resonances arises from the fact that they leave no observable tracks in bubble chambers. A particle moving with a velocity near that of light (about 10^{10} centimeters per second) leaves a trace long enough to be observed and measured even if it lives no longer than 10^{-10} second. But with a lifetime as short as 10^{-23} second it moves a distance no more than its own size (about 10^{-13} centimeter) before it ceases to exist.

The mass and all the other properties of resonances must therefore be inferred from the known properties of the particles which enter into their creation and result from their annihilation. A simple example of the way in which conservation laws aid in analyzing their interactions is afforded by the bombardment of protons by high-energy negative pions, resulting in the creation of a lambda resonance of 1520-Mev mass, plus a kaon:

$$p^+ + \pi^- \rightarrow \Lambda^{0*} + K^0$$

This is followed immediately by the decay of the resonance into a lambda baryon and a neutral pion:

$$\Lambda^{0*} \rightarrow \Lambda^0 + \pi^0$$

So far as can be observed, the two interactions occur at the same instant and at the same point. So there might seem to be but the single interaction:

$$p^+ + \pi^- \to \Lambda^0 + K^0 + \pi^0$$

The conclusion that two successive interactions are involved is based upon indirect but thoroughly convincing evidence involving the conservation of linear momentum (page 154). When a particle decays into *two* new particles, with release of energy from the decrease in mass, they fly apart with relative velocities which may be calculated easily and precisely from momentum considerations. If three particles are involved, however, the velocity relations are more complex and distinctly different. The analysis of the tracks actually observed in the bubble chamber proves that two two-particle interactions have occurred rather than one involving three.

As noted before, uncharged particles leave no visible tracks. These must be inferred from the visible tracks, if any, of their secondary decay products. The course of the Λ^0 may be found, for example, by locating the point at which it decays:

$$\Lambda^0 \to p^+ + \pi^-$$

By considerations of this sort, experimenters have been able to analyze phenomena occurring within intervals of space and time many trillions of times below the limits of direct observation. However, the analysis may become exceedingly involved. Only through the wedding of highly sophisticated electronic computers and ingenious detection equipment to high-energy accelerators has it become possible to understand the complicated experimental data of particle research.

SUMMARY OF PARTICLE EXPERIMENTS

IN about three decades after the discovery of the neutron the number of known kinds of particles, including the antiparticles, has increased from three to over eighty. This period has seen the development of new apparatus and new experimental techniques on a scale which the pioneers in particle research could not have imagined. No program devoted entirely to basic physical research, without applications to technology or industry, has ever achieved similar importance in terms of scientific manpower and financial resources.

One might ask why scientists have gone to such lengths to study entities which are so exceedingly rare in our familiar environment and have such a fleeting existence that their contribution to material substance, as we know it by direct experience, is utterly negligible. There is the immediate answer that all physical phenomena are the proper concern of science. The actual march of events has shown that particle research, though it has been concerned largely with the study of matter in its rarer and more evanescent forms, has profoundly altered all previous conceptions about the nature of all material substance. It has demonstrated that matter is not, in general, as solid and enduring as it has seemed to be. Rather, it must be thought of as energy in a highly concentrated state, prone to explode into massless particles flying off with the speed of light.

The ultimate constituents of matter, the "smallest parts" of the ancient philosophers, occur in great variety; indeed, the variety appears to be endless. These entities display a volatility of the most extreme sort, a dance of creation and annihilation. A few of them are stabilized by a chain of special circumstances, but even these may be prodded into acts of annihilation under proper circumstances. Antimatter, whose very existence had been quite unsuspected, has been found to have properties analogous but opposite to those of familiar matter. This discovery revealed a basic symmetry of the universe which had not been apparent with the electron-proton pair of particles. Neutrinos, first postulated in a tongue-in-cheek manner, have been found to be a major constituent of the universe and a dominant factor in its development.

All of the violent activity among particles is ordered and controlled by a number of conservation laws which may be stated in such simple terms that no mathematical sophistication is needed to grasp their import. Some of these were familiar to classical science; others have come out of the investigation of particle phenomena. These laws merely impose certain broad restrictions. Within their bounds the rule is: whatever can happen will happen. The typical quantum mechanical concept of probability does exert, however, its all-pervasive influence; some happenings are more probable and occur more frequently than others.

Conservation laws account for many general characteristics of particles: that spontaneous decay processes result in particles of lesser mass, that the lifetimes of different kinds of particles vary widely and that a few of them are stable. Because of the conservation laws of charge, spin, strangeness and isotopic spin, these properties can have only certain discrete quantized values. From the conservations of lepton and baryon number it follows that undetected neutrinos are created in many interactions involving electrons and muons, and more generally that particles and antiparticles are created together in equal numbers. Most importantly, conservation laws account for the existence of our material universe. Only because of them is matter constrained from exploding into nothing but photons and neutrinos.

Particle research has revealed the existence of two new kinds of forces, the strong and the weak, and has shown why they were unknown to classical physics. The conception of force has been broadened to encompass the cause of various kinds of particle interactions other than attraction and repulsion, the strength of a force being gauged by the rapidity with which it can produce its effects.

The division of all particles, except for the photons, into three major families, the leptons, mesons and baryons, has proven significant in various respects. The mesons stand apart from the rest in that their spins are integers, that none of them are stable and, most notably, that they may be created and annihilated freely in any numbers, a characteristic they share only with the photons. The leptons and barons have half-integer spins in common. Both the mesons and baryons respond to the strong force while the leptons do not.

None of these new insights could have been obtained from a study of the few stable particles alone, nor could the rare quality of this stability have been appreciated if the greater number of unstable ones had not become known.

The period just reviewed has been primarily one of great *experimental* progress. The huge particle accelerators, together with the elaborate auxiliary equipment for the detection and study of particles, have produced a veritable flood of new experimental data. The job of correlating and explaining all of these observations has proven extraordinarily difficult.

The present situation in particle physics may be likened to that prevailing in atomic physics before the time of Bohr. Experimenters had produced a profusion of observations concerning the chemical and optical behavior of atoms; and much of this knowledge had been organized, as in the chemical tables of elements and in the many orderly wavelength sequences recognized by the spectroscopists. But all of this systemization was empirical, without a background of basic theory, until Bohr proposed his conception of atomic structure and dynamics, later given sound theoretical support through the development of quantum mechanics.

In the study of particle phenomena great progress has been made in recognizing families and groups within families and in relating group structures to the various intrinsic properties of the individual particles. In addition to the three general families, theoretical studies have revealed more detailed relations and groupings. But a single comprehensive theory to account for all particle phenomena is yet to be developed.

The large and growing number of known kinds of elementary particles surely implies that all of them cannot be truly elementary and independent entities. But the relations between them are subtle and complex, and it has become increasingly apparent that these relations are not of the sort which may be easily visualized intuitively. Indeed, it will be seen in the following discussion that the theoretical study of particle phenomena has brought forth concepts even more remote from familiar experience than any which have been encountered heretofore.

12

Ideas and Theories

THE size of particles compares to that of atoms as atoms compare to the scale of things in the world of familiar objects; both involve roughly a hundred-thousand-fold ratio in magnitude. A tiny grain of sand, perhaps a thousandth (10^{-3}) centimeter across, behaves in every way like an object of the large-scale world. But a downward plunge to a hundred millionth (10^{-8}) centimeter leads to a realm in which everything existing in space and happening in time is a manifestation of changing patterns of matter waves. Things arrange themselves in sequences of discrete configurations, changes occur in abrupt quantum jumps and the pertinent laws of motion determine only the probabilities of events, not the individual events themselves. These profound changes in behavior are due primarily to differences in the relative size of objects and their de Broglie waves. Large objects are enormous compared to their associated waves; atoms and their waves are similar in size.

In the second downward plunge of minuteness, from a scale of 10^{-8} to one of 10^{-13} centimeter, a contrast of this sort does not exist. Here the matter waves are again comparable in size to the tiny regions in which particle events occur. Their radically new characteristics must be laid to other causes, in part to the change of scale itself. By quantum-mechanical principles the wave packets associated with events restricted to tiny regions of space must be constituted of very short matter waves; and because of the de Broglie relation between wavelength and momentum, this implies large values of velocity and energy and brief interaction times. Therefore, particle phenomena are necessarily rapid and violent, so violent that mass and energy interchange freely, and matter loses the stability it displays under less drastic conditions.

Atoms are a "half-way stopover" between the things of every-day experience and the weird realm of particles. They could still be treated to some extent in terms of familiar concepts. Thus, the Bohr atom model is frankly a mechanism operating in a familiar, albeit altered, manner. Particles are, however, conceptually more remote from atoms than are atoms from sticks and stones.

It is hardly surprising that attempts to extend the methods of quantum mechanics, so sucessful in dealing with atomic phe-nomena, to the realm of particles have met with difficulties. To make progress, it has been necessary to devise different methods of attack for various kinds of problems, for the properties of par-ticles, for their groupings, their interactions, and so forth. There exists, however, one generally recognized theoretical method of dealing with particle phenomena, the *quantum field theory*, which is adequate, in principle, to cope with all aspects of particle physics. As the name implies, it is concerned with the relations of quanta and fields.

Electric and magnetic fields have already been discussed briefly as regions in which charges experience electric and magnetic forces. To physicists in the mid-nineteenth century, fields had a more tangible meaning. They were assumed to be conditions of *strain* in an ether, a tenuous elastic "jelly" filling all space. Where there is a field, the ether jelly is under a strain of tension or com-pression, different from its normal relaxed state; and these strains were thought to produce the forces acting upon electric charges. There was also the luminiferous ether, possibly different from the electric and magnetic ethers which, when set into oscillation at one point, could transmit the oscillatory strains as a light wave.

Maxwell began the development of his monumental synthesis of electromagnetism and optics (page 48) by constructing an elaborate model of a mechanical ether, presumably capable of transmitting the various field effects. But after having built the

electromagnetic theory of light, in which light appears as a combination of oscillating electric and magnetic fields propagated
together through space, he saw that his mathematical equations
contained everything of importance. In the publication of his
studies *On a Dynamical Theory of the Electro-magnetic Field*
(1864), he presented only the mathematical theory with no mention of the ether model. Although he had thus made the ether
unnecessary, neither he nor his contemporaries thought of casting
it aside. Even up to the beginning of the twentieth century almost
all physicists continued to believe in the reality of the ether or at
least in the need of retaining it as an intuitive conception. But
in 1905, in his famous publication on the theory of relativity,
Einstein showed that the idea of an entity filling all space and
acting as a stationary reference, relative to which all motions
could be described in an absolute manner, is untenable, that only
the *relative* motions of objects have meaning. After the ether had
thus been abolished, the fields remained, like the grin of the van-
ished Cheshire cat.

Yet fields, in particular the traveling electromagnetic fields of
the light waves, still retained a measure of reality. These carried
energy and momentum and could cause electric charges to oscillate. Again, it was Einstein who robbed them of these trappings
of reality when, by postulating the photons, he relegated the light
waves to a mere ghostly existence as nothing more than mathematical abstractions determining the gross average propagation of
flocks of photons.

Quantum field theory has wrought a curious revival in the
status of fields. Although they are still largely mathematical conceptions, they have acquired strong overtones of reality. In fact,
this theory asserts that fields alone are real, that *they* are the substance of the universe, and that particles are merely the momentary manifestations of interacting fields.

The way in which particles are derived from fields is analogous
to the construction of atoms out of patterns of matter waves in
Schrödinger's original conception of wave mechanics. Here the
properties of atoms, and their interactions with each other and
with photons, are described in terms of the configurations and
changes of these wave patterns. Similarly, the solution of the
quantum field equations leads to quantized energy values which

manifest themselves with all the properties of particles. The activities of the fields seem particlelike because fields interact very abruptly and in very minute regions of space. Nevertheless, even avowed quantum field theorists are not above talking about "particles" as if there really were such things, a practice which will be adopted in continuing this discussion.

The ambitious program of explaining all properties of particles and all of their interactions in terms of fields has actually been successful only for three of them: the photons, electrons and positrons. This limited quantum field theory has the special name of *quantum electrodynamics*. It results from a union of classical electrodynamics and quantum theory, modified to be compatible with the principles of relativity. The three particles with which it deals are well suited to theoretical treatment because they are stable, their properties are well understood and they interact through the familiar electromagnetic force.

Quantum electrodynamics was developed around 1930, largely through the work of Paul Dirac. It yielded two important results: it showed that the electron has an *alter ego*, the positron, and it gave the electron its spin, a property which previously had to be added arbitrarily. When it was applied to the old problem of the fine structure of the hydrogen spectrum (the small differences between the observed wavelengths and those given by the Bohr theory), it produced improved values in good agreement with existing measures. However, in 1947 two experimenters, Willis Lamb and Robert Retherford, made highly precise measurements of the small differences in energy levels, using instead of photons the quanta of radio waves, which are more delicate probes of far lower energy. Their results, which showed distinct discrepancies from Dirac's theory, stimulated renewed theoretical efforts. Three men, Sin-Itiro Tomonaga of Tokyo University, Richard Feynman of the University of California and Julian Schwinger of Harvard, working independently, produced an improved theory which at long last gave precise agreement with experiment. For this work the three shared the 1965 Nobel Prize in physics.

The study of particles by the methods of quantum field theory was begun at a time when only a few were known. Since the field associated with a particle represents all of its properties, there had to be a distinct kind of field for each kind of particle; and as their number increased, so did the number of different fields, a complication which pleased no one. Actually, little further progress was made in the two decades following the success of quantum electrodynamics. Attempts to deal with the strongly interacting particles, the mesons and baryons, were frustrated by seemingly insurmountable mathematical difficulties. Still, the idea of developing a basic and comprehensive theory of particles continued to have strong appeal. In the mid-1960's the introduction of powerful new mathematical techniques has yielded results which indicate that this may yet be accomplished.

THE ELECTROSTATIC FIELD

THE interaction of the electromagnetic fields, whose energy is carried by photons, and the electron fields, which manifest themselves as electrons, is already familiar in the production of photons by the activity of atomic electrons. It is, however, not apparent how photons, which travel through space with the highest possible velocity, might be involved in *static* electric fields such as those which hold electrons close to the atomic nucleus.

Here a new concept is needed, that of *virtual photons*. Their existence is due in a remarkable, yet logical manner to the Heisenberg uncertainty principle. One form of this principle (page 99) asserts that the uncertainty ΔE in the energy possessed by a system and the uncertainty Δt in the time at which it has this energy are related by the formula:

$$\Delta E \times \Delta t \geqslant h/2\pi$$

Because of the relativistic correspondence between energy and mass, this relation applies as well to the uncertainty Δm in mass, which is $\Delta E/c^2$. Applied to an electron, this means physically that its mass does not maintain one precise value; rather, it *fluctuates*, the magnitude of the fluctuations being in inverse proportion to the time interval during which they persist. Electrons effect their mass or equivalent energy fluctuations by emitting photons, but these exist only on the sufferance of the uncertainty principle. When their time Δt is up, they must vanish. They cannot leave the electron permanently, carrying off energy, nor can they deliver energy to any detection device, including the human eye. It is impossible for them to be seen or detected; therefore they are called *virtual*, not real. Yet there is a warrant for their existence; theories in which they are postulated yield results in agreement with experimental observation. In the language of quantum field theory the interaction of electron and photon fields brings about a condition in which by permission of the uncertainty principle virtual photons are continually created and destroyed.

Virtual photons of greater energy exist for shorter times and travel shorter distances away from the electron before they are annihilated; those of lesser energy reach out farther. In fact, they travel a distance equal to the length of their associated waves (radio waves, light waves and others), which may vary over the whole range of values from zero to infinity. This swarm of virtual photons darting outward from the central electron in all directions constitutes the electric field surrounding the electron. Calculations based on this concept show that the field is strongest close by and drops off in inverse proportion to the square of the distance, in agreement with Coulomb's law of electric force (page 27). Virtual photons are the *quanta* of all electrostatic fields. For large charged objects they are so numerous that they produce a sensibly smooth and continuous effect, identical with the classical field.

Two electrically charged objects exchange virtual photons. This produes an *exchange force* between them, a result which follows directly from the principles of quantum electrodynamics, but which has unfortunately no analogy in classical physics and cannot be visualized in terms of familiar experience. The theory shows that the force between charges of like sign is one of repulsion, that for opposite signs it is an attraction, again in agreement with experiment.

There are, however, further complications. The virtual photons, produced by the electron, interact with the electron field to produce additional virtual electrons, which in turn yield virtual photons, and so on. Thus the theory, starting with one electron, ends up with an infinite number of them. Fortunately, the magnitudes of the successive steps in this sequence drop off rapidly so that after much effort the results of all this complex activity could be calculated very precisely.

This production of secondary virtual electrons manifests itself in the hydrogen atom as a slight alteration of energy levels. It was this effect which Tomonaga, Feynman and Schwinger succeeded in determining correctly.

For situations in which sufficient energy is made available, one of the virtual photons surrounding an electron may be "promoted" to a real one. This explains real photon emission when atoms release energy by making transitions to lower energy states.

This discussion implies that electrostatic fields are created by the activity of virtual photons. The point of view of field theory is rather the other way about, the photons being thought of merely as the way in which electric fields interact with electron fields. It is quite in order, however, to use either concept, depending on which is more appropriate to the problem at hand.

THE STRONG-FORCE FIELD

A FEW years after it had been found that atomic nuclei are built of protons and neutrons, Hideki Yukawa, working toward his Ph.D. at Osaka University, undertook a theoretical study of the force which binds nucleons together. The successful description of the electromagnetic force in terms of virtual photons suggested to him that the strong nuclear force might be accounted for in a similar manner.

It was known that this force does not decrease gradually toward zero with increasing distance; rather, its range ends abruptly at about 10^{-13} centimeter. Yukawa concluded that the virtual particles associated with the strong-force field should be all of one mass. Assuming that they dart out at velocities close to that of light, he could estimate that they exist for about 10^{-23} second; and from this value of Δt he calculated that their mass Δm, as given by the uncrtainty principle, is somewhat greater than two hundred electron masses. Since particles having a mass intermediate between the electron and proton were unheard of at the time this prediction was made, it was received with considerable skepticism.

The way in which Yukawa's prediction was verified has already been discussed (page 144). The pions discovered in cosmic-ray studies are the real particles, not the predicted virtual ones. As is true of photons, virtual pions may be promoted to the real state if sufficient energy is provided. In this manner pions are produced in considerable numbers in the violent collisions of protons or neutrons.

Further studies of the strong-force field have shown that its quanta include not only the three kinds of pions, but the other mesons, the kaons and eta particles, as well. Just as electrons are centers surrounded by virtual photons, so protons and neutrons, and all the other baryons, are to be pictured as centers of darting virtual mesons. A proton is constantly fluctuating between being just a proton and being a proton plus a neutral pion or a neutron plus a positive pion. Similarly, a neutron may be just a neutron or a neutron plus a neutral pion or a proton plus a negative pion. These fluctuations may be indicated thus:

$$p^+ \longleftrightarrow p^+ + \pi^0 \qquad n \longleftrightarrow n + \pi^0$$
$$p^+ \longleftrightarrow n + \pi^+ \qquad n \longleftrightarrow p^+ + \pi^-$$

The double-headed arrows imply that the interactions proceed in both directions. Similarly, an antineutron may be at times a negative antiproton plus a positive pion.

The neutron, in fact, must be in the form of a proton plus a negative pion a good part of the time, for it acts as if it were a tiny magnet. Since magnetic effects are produced only by moving electric charge, the neutron cannot be devoid of charge; rather, it must have equal amounts of both kinds spinning together about a common axis. The idea that both the proton and the neutron consist part of the time of central particles surrounded by charged pions is supported by experimental measurements of their magnetic effects, which are due mainly to the whirling pions. In the protons, where this whirling charge is *positive*, the magnet and the mechanical spin point in the *same* direction; in the neutron with its *negative* pions the two are *opposed*.

Direct evidence for the complex structure of protons and neutrons has been obtained through bombardment experiments with high-energy electrons (page 135). The proton experiments are carried out by bombarding ordinary hydrogen while the observations on neutrons are made with heavy hydrogen, whose atoms have nuclei which are proton-neutron pairs (since free neutrons in quantity are not available). From observations on the scattering of the bombarding electrons, it is possible to determine the distribution of electric charge within the bombarded particles. It is found that the pions have a range of about 10^{-13} centimeter, in agreement with Yukawa's theory. This theory gives only the *range* of the strong force and yields no information about its strength or details of its nature.

Attempts have been made to formulate a theory of the *weak-force* field, involving yet another kind of unknown virtual particle. All attempts to track down this *W-particle* experimentally have been unsuccessful. Finally, the gravitational field is thought to be mediated by virtual *gravitons* which, like photons, must be massless since the gravitational field, like the electrical field, has a long range. There is at present no expectation of observing real gravitons, for their creation in sensible amounts would require the violent agitation of huge masses. The particles related to the four kinds of fields are the only ones not constrained by number conservation laws; all four may be created and destroyed freely in any numbers.

Force fields consisting of darting virtual photons and mesons are again a radically new conception regarding the nature of matter. Material particles do not simply exist statically; they are centers of intense activity, of continual creation and annihilation. Every atom is a seat of such activity. In the nucleus there is a constant interplay of mesons, and the space around it is filled with swarms of virtual photons darting between the nucleus and the electrons.

ACTION AT A DISTANCE

QUANTUM field theory is, from one viewpoint, an attack on a problem of ancient origin, the problem of *action at a distance*. The natural philosophers of Aristotle's Lyceum may well have been puzzled to observe that a piece of rubbed amber exerts an attraction on bits of straw over a short intervening space, a phenomenon which eighteenth-century physicists would ascribe to the electric field in the vicinity of the charge on the amber. But they were no doubt more concerned with the analogous but more conspicuous observation of the downward pull experienced by all objects on the surface of the earth. Classical physics attributed this pull to the gravitational field which surrounds all pieces of matter but is of appreciable magnitude only near very large pieces such as the earth. To say that a stone held in the hand is pulled downward because it is in the earth's gravitational field, however, is merely puting a name to ignorance. It does not detract a whit from the mystery that the stone "feels" a pull with no visible or tangible agent acting upon it.

Isaac Newton, who formulated the law of action of the gravitational force, was well aware of this mystery. In one of his letters to the classical scholar and divine Richard Bentley he expressed himself thus:

> . . . that one body may act upon another at a distance through a vacuum without the mediation of anything else, by and through which their action and force may be conveyed from one to another, is to me so great an absurdity that, I believe, no man who has in philosophic matters a competent faculty of thinking could ever fall into it.

Newton saw clearly that his universal law of gravitation is a *description* not an *explanation*. The German philosopher and mathematician Baron Gottfried von Leibnitz (1646-1716), among others of Newton's contemporaries, criticized his work on this account, holding that his famous formula for the gravitational force ($F = Gm_1m_2/r^2$) is merely a rule of computation not worthy of being called a law of nature. It was compared adversely with existing "laws," with Aristotle's animistic explanation of the stone's fall as due to its "desire" to return to its "natural place" on the ground, and with Descartes's conception of the planets caught up in huge ether whirlpools carrying them on their orbits around the sun.

This unjust valuation of his work was repudiated in many of Newton's writings, as in the following passage from his *Optics*:

> To tell us that every species of thing is endow'd with an occult specific quality, by which it acts and produces manifest effects, is to tell us nothing. But to derive two or three general principles of motion from phenomena, and afterwards to tell us how the properties and actions of all corporeal things follow from these principles would be a very great step in philosophy, though the causes of those principles were not yet discovered.

Concerning his law of gravitation, which he discussed in the *Principia*, Newton made his position clear:

> I have not yet been able to discover the cause of these properties of gravity from phenomena, and I frame no hypotheses. . . . It is enough that gravity does really exist and acts according to the laws I have explained, and that it abundantly serves to account for all the motions of celestial bodies.

This quotation shows how thoroughly Newton espoused the experimental philosophy. He clearly expected that, if ever the "cause" of gravity is found, it will be deduced "from phenomena," that is, from experimental observations, and that in the meantime it is advisable to "frame no hypotheses."

The conception of fields of force as streams of virtual particles supplies the means "through which their action and force may be conveyed," which Newton so urgently demanded. It mitigates the problem of action at a distance, for with virtual photons producing the electric field, what happens *to* the electron happens *at* the electron.

Here is a lesson about the need for caution as to what "makes sense" in science. Nothing would seem more sensible than the observation that a stone tossed into the air falls back to earth; it would be surprising if the stone failed to do so. Yet upon closer study this simple event is seen to involve the metaphysical difficulties of action at a distance, difficulties which achieve a measure of intuitive resolution only in terms of the strange conception of virtual gravitons. This may serve as a warning that what passes for an understanding of simple things may well be no more than a tacit consensus to stop asking questions.

13

Symmetries

SYMMETRIES AND CONSERVATION LAWS

SYMMETRY in esthetics connotes a balance or harmony in the arrangement of parts about a center. In science the word has a broader meaning: a situation possesses symmetry if some of its aspects are indifferent to the occurrence of changes. A perfect cube is symmetrical in the sense that nothing about it is changed when it is given a quarter turn about its axis.

Empty space has certain important symmetries. It is *homogeneous*, that is, it is the same everywhere; nothing is altered by a change of location. Also, it is *isotropic*: its properties are the same in all directions. These symmetries lead to corresponding *invariance principles*. Because of the homogeneity of space all physical phenomena are invariant to changes of location. The results of identical experiments done in London, Tokyo and New York are identical. Similarly, the isotropic nature of space makes happenings invariant to their orientation. But for these invariance principles the conduct of affairs on our earth, moving through space and rotating on its axis, could be highly complicated.

There is an analogous symmetry to time; no one particular time differs of itself from any other. This results in time invariance. The course of events under identical conditions is the same whether they happen yesterday, today or tomorrow.

Invariance principles are of great importance because they generate *conservation laws*. If happenings were not independent of

184

location, an object moving freely through empty space might
change its velocity merely because it has entered a new region.
This does not happen, however; the object moves with constant
velocity and constant momentum, that is, the law of conservation
of linear momentum holds. Similarly, a rotating object does not
change its state of rotation as it turns to new orientations in space,
thus conserving its angular momentum. Time invariance leads
to conservation of energy, or rather mass-energy. If changes could
occur merely because of the passage of time itself, energy obviously would not necessarily be conserved.

Yet another symmetry of space is that of *reflection*. The reflection symmetry considered in the mathematical description of
space is of a more thorough sort than that of a plane mirror.
Whereas a mirror reverses only one of the three rectangular directions of space, reflection in the scientific sense reverses all three,
left to right, front to back and top to bottom. The reflection symmetry of space implies that the reflection of any phenomenon
proceeds in a manner identical with that of the phenomenon
itself, except that all directions are reversed. Expressed more
simply, this symmetry principle means that the reflection of an
event as seen in a mirror looks as real and possible as the event
itself. Obviously, this principle is in accord with all ordinary
experience.

From the invariance of phenomena under space reflection there
follows the law of *conservation of parity*. Parity is yet another
significant property of particles, as are spin, strangeness and the
rest (page 161) whose conservation laws have already been considered. Its definition for particles derives from their association
with patterns of matter waves in quantum mechanics. If the
nature of the wave pattern is such that it remains unchanged
under space reflection, the particle is said to have *even* parity
(designated as $p = +1$); if reflection changes the sign of the
pattern everywhere, the parity is *odd* $(p = -1)$. Because the
physical nature of any situation is determined by the *square* of
the wave pattern values, it follows that space reflection symmetry holds for odd as well as for even parity, for the squares
of -1 and of $+1$ are both equal to $+1$. Therefore, a mere change
of sign at all points of the wave pattern does not alter the physical
situation to which the pattern pertains.

The law of conservation of parity states that in any particle interaction the total parity remains unchanged. It is the same for the particles resulting from, as for those entering, an interaction. In accord with the principles of quantum mechanics this conservation law necessarily implies invariance under space reflection; and it results further in the general principle that there can be no phenomena in nature which make a fundamental distinction between right and left, or more generally, between any given direction and its opposite. Any process which produces such an essential asymmetry causes a change in parity, that is, it violates reflection symmetry.

To understand why this is true, assume that there is a law of nature concerning certain kinds of spinning tops which states that they emit sparks equally upward and downward, as seen at the left of Figure 29 both when spinning in the right-hand and the left-hand direction. In this situation there is no fundamental distinction between up and down or between right- and left-hand rotation. Thus, the mirror image of a right-hand spinning top, showing a left-hand spinning one emitting sparks both at the top and bottom, is a possible event in accord with space reflection symmetry. If there were tops which do make a distinction between opposite directions, which may emit sparks upward only

Figure 29. Spinning tops and their mirror images.

when spinning to the right and downward only when spinning to the left, the mirror image of a right-hand spinning one, showing a top spinning to the left and emitting sparks upward, would be an impossible event. For a top which shows a preference for any one particular direction, reflection symmetry does not hold. This is an example of the above general principle. Any phenomenon which is restricted to act predominantly in one particular direction is a violation of the conservation of parity. This conservation law is thus linked to the reflection symmetry of space as the law of conservation of spin follows from its isotropic symmetry.

The intimate relation of all these various conservation laws to the invariance principles of space and time gives them a special aura of infallibility. The news that one of them had been violated caused a great stir in the community of physics.

THE NONCONSERVATION OF PARITY

THE law of conservation of parity served, like the other conservation laws, as a useful guide in the analysis of particle interactions. Its validity was never in doubt until there arose, in 1955, the *tau-theta paradox*. Two mesons, called tau and theta, had been found to be identical in mass and in all other properties so that they could be considered as the *same* particle. It was observed, however, that the tau particle decays to three pions which together have odd parity while theta changes into two pions with even parity. In accord with parity conservation it was necessary to assign odd parity to tau and even to theta, implying that they are two *different* particles.

The theoretical physicists Tsung Dao Lee of Columbia University and Chen Ning Yang of the Institute of Advanced Study at Princeton became interested in this puzzle. They noted that the tau and theta decay processes violate the conservation of strangeness and must therefore occur via the weak force. If it were shown that weak-force interaction may also violate the conservation of parity, tau and theta could be one and the same particle. (Both are now known to be the positive kaon.) Following up this idea, they investigated other known weak-force interactions and found that there was actually no good evidence to prove that parity *is* conserved in any of them. So certain had physicists been that parity is conserved "of course" that this had simply been overlooked. Once parity conservation had been questioned, it had to be tested.

The crucial experiment was performed in 1956 by Chien-Shiung Wu, Professor of Physics at Columbia University. She had conducted important experiments in support of Fermi's theory of weak interactions (in which he had predicted the neutrino) and had been cited as "the world's foremost female experimental physicist" when receiving an honorary doctorate of philosophy at Princeton University.

The principle involved in Dr. Wu's experiment is analogous to that given for the spinning and sparking tops. The "tops" in the actual experiment were the spinning nuclei of radioactive cobalt atoms, and the "sparks" were the beta rays (electrons) which they emit. This experiment was chosen because beta-ray emission is a weak-force phenomenon. Since each cobalt atom emits but one electron as it undergoes radioactive decay, it was essential to line up the spins of all of them in the same direction so that the very large number of atoms in the sample of cobalt would all emit the electrons in reference to this common direction. This was accomplished by placing the cobalt in a strong magnetic field, which lines up its magnetic atoms like compass needles. It was necessary to cool the cobalt very close to the absolute zero of temperature, where random thermal motion of atoms ceases; otherwise this motion would have disrupted the magnetic alignment. In the very difficult technique of obtaining such extremely low temperatures, Dr. Wu had the assistance of a team of scientists at the U.S. Bureau of Standards, where the experiment was performed.

If parity were conserved, if nature showed no preference for right or left, equal numbers of beta rays would be emitted in both directions relative to the spin of the cobalt atoms. The experiment showed that the emission is predominantly in one direction, thus proving that space reflection symmetry does not hold for weak-force interactions. (No violation of parity is found in phenomena resulting from the action of the electromagnetic or the strong force.)

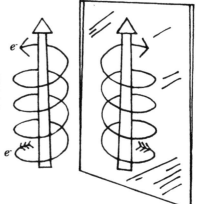

Figure 30. The preferred direction of beta-ray emission from cobalt nuclei in the magnetic field produced by electrons flowing through a wire helix.

In the analogy of the spinning tops this outcome is equivalent to the discovery that there *is* a law of nature which says that left-hand spinning tops emitting sparks upward do not exist. This discovery was reported promptly, not only in the scientific press, but also in the public journals throughout the world.

Returning to the experiment of Dr. Wu, the helix in Figure 30 represents the coil of current-carrying wire which produced the magnetic field. The direction in which the electrons flow through the wire is indicated; the large arrow shows the preferred direction of beta-ray emission. (There must of course be wires, not shown, connecting the ends of the helix to a source of current.) The relation of directions is a right-hand one: if the fingers of the right hand are curled in the direction of the electron flow, the extended thumb points in the preferred direction of emission. The mirror image shows a left-hand relation, which is not a possible one.

If nature did not exhibit this fundamental nonconservation of parity, it would be impossible for the earth's inhabitants, if they were to establish radio communication with intelligent beings on some distant planet, to convey to them our meaning of left and right or, what comes to the same thing, of clockwise and counter-clockwise rotation. It would not do to refer to clocks or to the location of the heart for all these might be the other way about on planet X. The inhabitants could be given instructions, however, on how to produce a magnetic field by winding a wire helix and sending an electric current through it. They could be instructed further on how to set up the radioactive cobalt experiment; and then could be told that, if they look through the helix in the direction in which most of the beta rays are emitted, the electron stream will be traveling in a way which is called clockwise on earth.

This would do for the inhabitants on a planet in our own galaxy, perhaps a thousand light years distant. Beings in other galaxies, millions of light years away, might be made of antimatter; and here the description of right and left would fail. To see why this is true, the mirror reflecting the wire helix may be thought of as having the power, not only to reverse direction in space, but also to change particles into antiparticles. The left-hand electric current in the mirror helix is then a stream of *positrons*, which has exactly the same magnetic effect as the right-hand current of negative electrons in the real world, so that the mirror image is again a possible phenomenon. (There are actually two additional changes: the nuclei of the anticobalt atoms are made of antinucleons, and they emit beta rays which are antielectrons, but these two circumstances annul each other.) Although parity conservation is violated, a broader symmetry involving both space reflection and particle-antiparticle exchange is still conserved.

An experiment performed in 1966 does seem to afford the possibility, in principle, of determining whether our extragalactic correspondents consist of matter or antimatter. A team of physicists at the Brookhaven National Laboratory found that, in the decay of the eta meson into three pions, one positive, one negative and one neutral, the positive one flies off with a distinctly higher energy than the negative. Since the neutral eta particle is its own antiparticle, physicists in an antiworld would be performing this experiment with the same particle used in our world. No matter how they determined energies, the pion whose charge we call positive would show the greater value. If they reported that the charge of the particle having greater energy is opposite in sign to that of their atomic nuclei, it would be evident that theirs is an antiworld. Unfortunately, the two-way conversation by radio waves traveling at the ultimate velocity would require millions of years to complete.

However, when the eta-decay experiment was repeated at CERN, the asymmetry between the two kinds of pions was not observed so that its reality is still in doubt.

Light and radio signals, which originated in distant stars many millions of years ago, are reaching our earth today. There is no known way of determining whether they arose in matter or in antimatter since photons are their own antiparticles. Antineutrinos are, however, distinguishable from neutrinos so that, as was suggested previously (page 147), neutron astronomy does afford a possibility of detecting antigalaxies.

Relations of conservation laws to symmetries are of particular interest in particle physics because there are additional conservation laws, of baryon number, strangeness and other particle properties, which still stand unexplained and unrelated to any basic principles. It is possible that these conservation laws may be associated with as yet unknown symmetries existing in abstract mathematic spaces different from the familiar space of everyday affairs. Mathematicians are familiar with various kinds of spaces besides the one which they refer to as "ordinary three-space." These may have many more dimensions than three, and they may be made up of various kinds of entities other than mere distances. A simple example is a space of six dimensions, three of position and three of velocity, in which a particular location determines completely both the position and the motion of a point. Such mathematical constructs have proven powerful tools in dealing with scientific problems.

GROUP THEORY

THE word *group*, as used in mathematics, refers to a set of entities which have certain simple but well-defined properties. One of these is that the members or *elements* of a group must be of such a nature that the product of any two of them is also an element. The positive integers, 1, 2, 3, 4, 5 . . . have this property $(2 \times 3 = 6$, etc.). They do not satisfy, however, two further rules for groups: there must be a *unity element* which leaves all elements unchanged in multiplication; and to each element there must be an *inverse element* such that the product of the element and its inverse is unity. The positive integers, together with all the rational fractions (fractions made by dividing one integer by another, as 6/5, 27/53, etc.), form a complete group. Its unity element is simply *1*, and the inverse element of 7 is 1/7, of 21/15 is 15/21, etc. All of this seems too simple and obvious to be of much value. Yet the concept of groups has proven a valuable tool in dealing with the complex properties of particles and the relations among them.

Group theory was developed throughout the nineteenth century as a branch of "pure" mathematics, with no thought of applications to the real world. It deals with the many possible kinds of special groups and with the mathematical operations which may be performed with them. The properties of groups are stated in general terms, and they hold no matter what the group might represent physically. (To give a simple analogy: the statement that $2 \times 2 = 4$ gives a relation of abstract numbers which holds for a great variety of real things.) Group theory is at a higher level of abstraction. Over nearly a century of development it progressed from dealing with simple numbers to become a way of manipulating abstract mathematical symbols without reference to their physical connotations. Therefore, it is possible to carry over the laws and techniques of group theory to situations where, as is frequently the case in particle physics, the meaning of the elements cannot be readily visualized. Group theory may serve as a guide where intuition is blind. Given a properly chosen group, its elements actually become the various properties of particles, some of which, such as charge and spin, have tangible physical meaning while others, strangeness, isotopic spin, and so on, are obscure.

Of particular significance are the kinds of groups which deal with *symmetries* and in which the elements are *operations*, for example, the group of rotations of the cube. The elements of this group operate on the cube so as to bring its corners into positions formerly occupied by other corners. A quarter turn about an axis through the center of two opposite faces is an element of this group and so is a half or a three-quarter turn. (An eighth turn is not.) The product of elements in this group means two turns in succession; thus, the product of a quarter and a half turn is a three-quarter turn, which is also an element. The unity element

Rest mass, m	Strange-ness, S	Hyper-charge, Y	Iso-topic spin, I	Particle				Mass differences, Δm
1676	−3	−2	0	Ω^-				⎫ 146
1530	−2	−1	1/2	Ξ^-	Ξ^0			⎬ 145
1385	−1	0	1	Σ^-	Σ^0	Σ^+		
1238	0	+1	3/2	Δ^-	Δ^0	Δ^+	Δ^{++}	⎬ 147

Figure 31. Chart of particles comprising an SU(3) group.

is no turn at all, and the inverse element is a turn of the same magnitude the other way about. The two examples, the group of numbers and of rotations, are indicative of the great variety of entities which may have the properties of a group.

One particular group which has revealed a large measure of order in the complex array of baryons is the *special unitary group,* designated as SU(3). The importance of this group was discerned in 1961 by Murray Gell-Mann, professor of physics at the California Institute of Technology, and independently by Yuval Ne'eman at Tel Aviv University. They found regularities among the eight oldest known baryons (the nucleons and the lambda, sigma and xi baryons) corresponding to the properties of this group.

Of greater interest are the ten baryon resonances, shown in Figure 31. All of these particles have a spin of 3/2 and positive parity and differ in an obviously systematic manner in their values of strangeness S, hypercharge Y and isotopic spin I, and notably in their mass values as seen in the nearly constant mass differences Δm. (The given mass values are the averages over slight differences for the particles of various charges in each kind.)

Only the deltas and sigmas were known when Gell-Mann and Ne'eman proposed the SU(3). After the discovery in the following year of the two xi resonances, they were encouraged to predict the existence of the tenth missing particle, which they named the *omega.* Since they had also succeeded in predicting its properties, they could suggest the proper experimental procedure to create it. Because of its unique strangeness value ($S = -3$) it would have to be created in association with several kaons ($S = +1$) to conserve strangeness; this meant that a large amount of energy would be required. However, by bombarding protons with anti-kaons, only two kaons need to be created:

$$p^+ + \overline{K^-} \rightarrow \Omega^- + K^+ + K^0$$
$$S = 0 \quad\ -1 \quad\ -3 \quad +1 \quad +1$$

After a month of round-the-clock labor during which about fifty thousand bubble chamber photographs were scanned, a team of scientists, working with the accelerator at Brookhaven, found a bubble track which could be identified as having been made by the omega-minus particle. Because of its short lifetime, less than 10^{-10} second, its track was only a few centimeters long; and the identification had to be made by a study of the decay products which are produced in a sequence of several steps:

$$\Omega^- \rightarrow \Xi^0 + \pi^-$$
$$\Xi^0 \rightarrow \Lambda^0 + \gamma + \gamma$$
$$\Lambda^0 \rightarrow p^+ + \pi^-$$

with violation of strangeness at each step.

The discovery of this particle was of particular importance, for it occurred at the time when the "official" particle theory, the quantum field theory, was proving incapable of dealing with the baryons. Group theory seemed the more promising of several alternative methods, and its successful extension to the newly investigated resonances established it as a procedure which particle theorists might pursue to advange.

Further studies have revealed even larger group relations. Eight baryons with spin 1/2 and ten resonances with spin 3/2 can be combined into one more complex group, the SU(6). Similar group relations have been developed among the mesons. Here the groups include both particles and antiparticles, whereas these two fall into separate groups for the baryons.

In keeping with their isolation from the other kinds of particles, the leptons show no group characteristics. Failure to find any new leptons in many years of search implies that there are no others. For the theorists the muon is already one too many; its existence is quite inexplicable.

QUARKS

CLOSELY related to group theory is the concept of *quarks*. This word was coined in an antic mood by the writer James Joyce in *Finnegans Wake*: "Three quarks for Muster Mark. . . ." This seemed to Gell-Mann, who introduced it into particle physics in 1964, to be a suitable name for something which comes in threes and which might exist but probably does not.

Quarks are hypothetical, as protons and neutrons would have been if the periodic table of elements had started with helium and free nucleons had never existed. The obvious regularities in the mass and electric charge values of the various nuclei certainly would have suggested that they might be constituted of two kinds of particles of equal mass, one positively charged, the other neutral. Some physicists might have decided that these entities are no more than a convenient fiction while others could have taken them seriously and proceeded to discover them. Similarly, the baryons and mesons might be thought of as being made up of simpler entities, the quarks. Most physicists consider them merely a convenient scheme for organizing the various particle properties. Some, however, suspect that quarks might have independent physical existence; and they have carried out experiments designed to discover them.

In this scheme of organization each kind of baryon is a group of three quarks while each meson is a quark-antiquark pair. The individual quarks are given such values for their various properties so that, by suitable grouping, the correct properties for all the kinds of particles are obtained. For example, all quarks are given a baryon number of $+1/3$ while for antiquarks the value is $-1/3$. (All properties of antiquarks are opposite in sign to those of quarks.) This gives the proper value of $+1$ for the baryons and zero for the mesons.

Table 5 shows the values of baryon number B, electric charge Q (in electron units), strangeness S, and the "vertical" component of isotopic spin I_3 (page 164) for the three kinds of quarks designated as p, n and λ. It also lists a few meson and baryons together with their quark structure and the resulting values of their properties. (The proton and neutron are given capital letters to distinguish them from the quarks.) The spins of both quarks and

Table 5. Table of quarks.

Particle	Quark struc- ture	Baryon num- ber, B	Electric charge, Q	Strange- ness, S	"Vertical" compo- nent, I_3	Rest mass, m (Mev)
	p	1/3	2/3	0	1/2	?
	n	1/3	−1/3	0	−1/2	?
	λ	1/3	−1/3	−1	0	?
π^-	$\bar{p}n$	0	−1	0	−1	140
K°	$n\bar{\lambda}$	0	0	1	−1/2	498
$P+$	ppn	1	1	0	1/2	938
N°	pnn	1	0	0	−1/2	940
Ξ°	$p\lambda\lambda$	1	0	−2	1/2	1192
Δ^-	nnn	1	−1	0	−3/2	1242
Δ°	pnn	1	0	0	−1/2	1239
Δ^+	ppn	1	1	0	1/2	1236
Δ^{++}	ppp	1	2	0	3/2	1234
Δ^-	nnn	1	−1	0	−3/2	1242
Σ^-	$nn\lambda$	1	−1	−1	−1	1385
Ξ^-	$n\lambda\lambda$	1	−1	−2	−1/2	1530
Ω^-	$\lambda\lambda\lambda$	1	−1	−3	0	1676

antiquarks are 1/2. By twos they can orient themselves so as to give values of zero and 1 for the mesons and their resonances; in threes, they give spins of 1/2 for the baryons and 3/2 for their resonances. (Still higher values of spin are due to more complex associations of quarks.)

The last eight particles in the table are the baryon resonances shown in the SU(3) group (page 193). The four deltas have nearly the same mass; but through the sequence of the last four there is a considerable increase of mass at each step. This seems to imply that the n and p quarks are of nearly equal mass while the λ is more massive. In the quark scheme as a whole there is no correlation between mass and quark composition. The neutron and the neutral delta, for example, are both *pnn* combination. Yet their masses are, respectively, 940 and 1239 Mev.

This is not surprising in view of the typically labile behavior of mass in particle phenomena. If quarks actually exist, they must be very firmly bound as particles; and the great amount of energy required to separate them would make them far more massive in the separated state. Thus, with the same three quarks, the particle in which they are more loosely bound contains the large amount of energy needed to effect this looser structure and has therefore the greater mass.

It has been estimated that each one of the three quarks composing the proton, for example, must have a mass at least five times as great as that of the whole proton. This follows from the observation that the energy available for bombarding protons in the most energetic accelerators now in use is evidently not sufficient to blast them apart into quarks. The possibility that higher energies might produce free quarks is one of the motivations for building still larger machines. (Among others is the desire to learn more about the nature of the nuclear force and to study the elusive neutrino, although, judging by past experience, the most important new findings may well be of a new kind not yet envisioned.)

It might seems that quarks are the answer to the long quest for a small number of more elementary entities of which the "elementary" particles are constituted. This simplified discussion has, however, neglected certain complexities. Thus, for good theoretical reason the quark structure of the neutral pion is given as $\bar{p}p - \bar{n}n/\sqrt{2}$, which makes a naïve picture of particles as simple groups of quarks difficult to maintain. Also to be kept in mind is the fact that none of the leptons can be represented in this manner.

14

More Theories

IN the 1950's it was becoming increasingly evident that quantum field theory could not cope with the problem of strong-force interactions among the mesons and baryons. Considering this situation, Heisenberg sensed an analogy with the failure of the Bohr theory to give a logical account of atomic phenomena. In this field he had developed a successful dynamical theory, matrix mechanics, by excluding from consideration the unobservable electron orbit transitions (page 81). Quantum field theory, he realized, was again dealing with unobservables, for fields are but mental constructs, and it is the particles that are actually observed.

In the language of particles the absorption and emission of radiant energy by hydrogen atoms is an interaction between protons, electrons and photons. What Heisenberg had done in matrix mechanics was to consider the state of these particles before and after the interaction while ignoring the unobservable details of the interaction itself. For the hydrogen atoms, which involve three well-known stable particles interacting by the familiar electric force, this approach had been highly successful. He undertook to build a general theory of particle interactions based on the observable properties of the intial and the resulting particles without the need of following in detail what happens in the brief instant of collision.

Scattering is the general term among particle experimenters for

198

everything that is observed in particle collisions. Since Heisenberg's theory deals with these phenomena by the mathematical theory of matrices, it is known as scattering-matrix or *S-matrix theory*. The S-matrix itself is an array of symbols, arranged in rows and columns (page 82), which represent the transitions from the original to the final particles in a given interaction. The matrix is said to *operate* mathematically to effect the transitions. The symbols or elements are made up by considering the properties of the original and of all possible final particles in accord with the mathematical rules of matrix theory.

What actually happens when one particle meets another with a certain amount of energy can be discovered by experiment. Furthermore, conservation laws determine what interactions are possible, and a knowledge of the forces involved shows why some interactions are faster than others. The object of S-matrix theory is to provide a general and basic description of these matters and to account for further details, such as the relative probabilities or frequency of occurrence of the various possibilities and minor differences in interaction rates.

The goals of the S-matrix method are less ambitious than those of quantum field theory. Its scope is restricted to particle *interactions*; thus, it supplements group theory, which deals with particle *properties*. It remains close to facts of observation and uses the relatively simple and well-known mathematics pertaining to certain special kinds of symmetrical matrices. While it falls far short of being a complete dynamical theory of particle interactions, it has revealed interesting relations among observed facts and has made useful predictions about others still to be observed.

FEYNMAN DIAGRAMS

DURING the development of quantum electrodynamics in 1947-49, Tomonaga and Schwinger had held close to the classical conception of fields as real entities while Feynman, though he arrived at substantially the same results, had given more emphasis to their particle aspects. Continuing independently in this bent, he developed a very general and successful method of dealing with particle interactions. This was worked out by adopting ideas both from quantum field theory and from the S-matrix method while improving on the latter in two important respects. He developed powerful new mathematical methods, the *Feynman integrals*, for describing particle transformations; and he devised an extremely simple and useful graphical technique, the *Feynman diagrams*, through which particle interactions may be displayed in a lucid pictorial manner.

To understand these diagrams, it is necessary to comprehend the meaning of the "four-dimensional space-time continuum," a concept developed by the Russian mathematician H. Minkowski (1864-1909). He showed that the mathematical formulas of relativity theory could be expressed more simply and naturally if the three dimensions of space, length, width and depth, and the one extension of time are combined into a single, indivisible space-time. The trajectory of a point moving in this space-time is called a *world line* to distinguish it from a *path* in space alone.

Four dimensions are impossible to present diagrammatically. But the essential ideas of space-time motion may be illustrated on a two-dimensional graph in which displacements to the left and right represent motion in space while those in the upward and downward directions indicate progress through time, *up* into the future and *down* into the past. A line in this space-time plane is the world line of a point which is moving through space in one direction only.

It is impossible for an object to "stand still" in the space-time plane, for if it is at rest in space, it is moving directly upward in time; its world line is vertical. If it moves to the right, its world line is inclined in this direction, the more so if it moves faster.

Figure 32 shows the world line of an airplane on a trip due east from New York to Naples. From points A to B the airplane, and the city of New York, are moving together in the vertical time direction. At the moment B the airplane takes off, traveling eastward in space and upward in time until at C it meets Naples, which has been moving upward in the meantime. After this moment, at which New York has reached the point D, both cities and the airplane continue to move upward at the same rate.

The Feynman diagrams of Figure 33 are the world lines of interacting particles. The arrows, labeled with the symbols of the

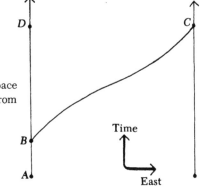

Figure 32. World line, through space and time, of an airplane flying from New York to Naples.

particles, indicate the direction of travel through space-time. Diagram (1) indicates the interchange of a virtual photon between two electrons, resulting in mutual repulsion. (Actually, two approaching electrons exchange a large number of virtual photons.)

At (2) is shown the exchange of a virtual pion between a proton and a neutron. This not only produces an attraction but changes each particle into the other. However, the naïve description of this event as "the transfer of a virtual pion from the proton to the neutron" is not to be taken literally. What is actually observable is that the proton becomes a neutron and the neutron a proton. A more detailed description is unwarranted, for the virtual pion, although it is the basis for a useful theory of the nuclear force, is unobservable even in principle.

The more complex diagram at (3) shows the train of events which led to the discovery of the electron neutrino or rather, as it turned out later, the antineutrino. A proton at rest is struck by a high-energy antineutrino resulting in an interaction which produces a neutron and a positron (page 154). The latter immediately meets an electron and by mutual annihilation gives rise to a pair of photons, each of 0.51-Mev. energy. A few millionths of a second later the neutron is captured by a cadmium nucleus, initiating a nuclear reaction (not shown) which produces several photons with a total energy of 8.9 Mev. All of these photons are

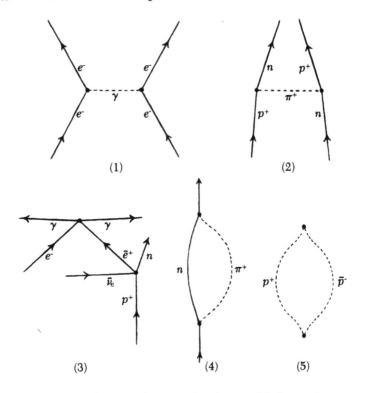

Figure 33. Feynman diagrams of various particle interactions.

detected and their energy is measured. Since no other conceivable interaction can give this particular sequence of photon emission, these measurements prove that a rare neutrino capture has occurred.

The process which Yukawa postulated to account for the strong force field is shown at (4). A proton becomes, for the brief time permitted by the Heisenberg uncertainty principle, a neutron plus a positive pion. Most surprising of all perhaps is the event shown at (5), for there is no world line at all leading up to it or away from it. A proton-antiproton pair apparently appears out of nowhere, exists briefly and then vanishes again. Yet even this violation of mass-energy conservation is permitted if it is of sufficiently brief duration, the order of 10^{-24} second.

According to the principles of quantum field theory, particles are associated with fields extending throughout space, which means that fields exist even where there are no permanent particles, that is, where there is a *vacuum*. From these principles it follows that a vacuum is not an empty space. Rather, it is a seat of continuous activity with virtual particles of many kinds winking in and out of existence. Physicists speak of a *physical vacuum* as distinct from the *bare vacuum* of classical physics. Although these vacuum phenomena violate mass-energy conservation, they are in accord with the many other conservation laws of charge, spin, linear momentum, baryon number and all the rest.

Feynman diagrams suggest a surprising but useful idea concerning the nature of antiparticles. To fix ideas, it will be assumed that the diagram at (1) of Figure 34, showing electrons and antielectrons, is covered by a card in which there is a narrow slit, indicated by the dashed lines. With the passage of time the card moves upward, and the observer "sees" through the slit the state of affairs at successive moments. As time progresses from the instant

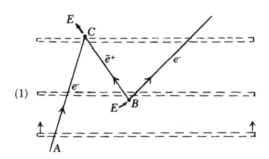

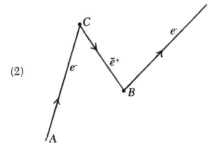

Figure 34. Feynman diagrams of the creation and annihilation of an electron-positron pair.

A toward B, he sees a single electron moving to the right. At the instant B an electron-positron pair is created, through some process not specified, which supplies the necessary energy E. At C the positron and the original electron meet in mutual annihilation with release of the energy E, leaving only the newly created electron moving off to the right. This is the train of events as seen by an observer who is aware, in the usual manner, of events occurring in sequence with the passage of time.

If the card is removed to reveal the whole sequence of events at a glance, a simpler description is possible, as shown in the diagram at (2). An electron travels to the right during the time interval from A to C and then changes into a positron. It continues as a positron up to the instant B, at which it changes back into an electron. This is obviously a simpler story than the previous one involving three particles. But it does introduce the weird notion of a particle moving *backward* in time. Feyman found that the mathematical analysis is also simpler in the second interpretation of the events. In general, he could deal easily with antiparticles by treating them as particles moving opposite to the familiar way in time. How much physical reality might be attached to this conception (whether beings in an antiworld live "backward") is an intriguing question.

PARTICLE ENERGY LEVELS

It is logical to consider an atom in its normal state and in an excited state of higher energy as two distinct "particles." Because of its greater energy content the latter has more mass (though the increase is too small to be measurable), and it also has a different size and angular momentum. This point of view suggests that certain groups of particles having properties in common might be the "ground" state plus a sequence of "excited" states of the same particle.

There are, however, a number of important differences as well as analogies between the two situations. The quantum-mechanical laws which govern the dynamics of atoms are well understood so that it is possible to determine the properties of their various energy states. But the nature of the strong force and the laws of its action are still unknown; therefore, similar calculations cannot be made for the excited states of particles, although it is evident that the energy levels must differ by millions of electron-volts (as compared to a few electron-volts in atoms) and must correspond to large changes of mass.

When an atom has been raised to an excited state, it returns to the ground state through emission of photons, which are the quanta of the electromagnetic field. Similarly, an excited baryon returns to its ground state by emitting mesons, the quanta of the strong-force field. Furthermore, the transitions between energy levels of particles frequently involve changes of electric charge and of strangeness as well. Those in which there is a change of energy only may be effected by neutral pions. Where a change of charge occurs, one or more charged pions are required. If there is, in addition, a change in strangeness, kaons and antikaons are involved. A few examples of such transitions are shown in Figure 35, which is obviously analogous to the hydrogen energy-level diagram constructed by Bohr (page 58). (Resonances are indicated by asterisks.)

Figure 35. Energy-level diagram for particle transformations.

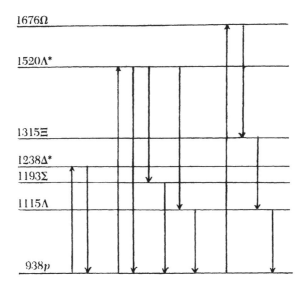

The transition from the proton to the delta resonance, both of zero strangeness, is effected by pion absorption; and in the return to the proton state pions are again emitted. After the upward transition to the lambda resonance the return may be in one step or in two, either via the sigma or the lambda. The jump from the proton to the omega level with a change of strangeness from zero to −3, is brought about by a high-energy antikaon. Here there is an additional complication, for to conserve strangeness kaons are *emitted* in the upward jump. The return to the ground state proceeds in three steps, from omega to xi to lambda to proton. It is mediated by the weak force with an increase of one unit in strangeness for each step.

As seen in the diagram, all the upward transitions are accomplished in a single step, starting from the lowest level. This is due to experimental conditions. Because of the extremely short lifetimes of all the excited-state particles, it is impossible to produce a concentration of them sufficiently large to serve as an adequate target for bombardment to higher levels.

A similar scheme of energy levels has been devised for the mesons, with the pion, kaon and the eta particle as the ground states. Here the excited states are meson resonances, and the transitions to lower states are effected by the emission of pions and kaons. The pion resonance of 750 Mev, for example, decays into two ordinary pions; this is interpreted as a transition of the resonance to the normal pion state with the emission of a second pion.

Although it is not possible to calculate the mass values of the excited states or the transition probabilities between them, there is a conceivable mechanism through which energy absorption or emission might occur. In terms of the virtual particle concept of strong-force fields, both meson and baryons are complex structures consisting of a center surrounded by virtual mesons. Transitions between excited states of particles could be thought of as changes in this virtual particle structure analogous to the changes in electron structure occurring in atoms during energy-level jumps.

The idea that the hadrons (the mesons, the baryons and their resonances) are related in the manner of different energy states serves to emphasize the similarities among them implied by other methods of analysis, and by the fact that interactions between all of them are brought about by one and the same force, the strong force, which quantum field theory explains in terms of virtual mesons. In addition, there is the obvious similarity in their masses. There are no really large ratios among them; the greatest mass is less than twenty times the least. All of these considerations have been expressed in the conception of *particle democracy*, no one particle being thought of as fundamentally different from all the rest. Even the unique stability of the proton is due to the circumstance that it is the lightest of the baryons, a quantitative but hardly a qualitative distinction in kind.

These relations are clearly reciprocal: the proton, for example, is a lower energy state of the delta just as the delta is a higher state of the proton. Thus, the proton must have within itself the potentiality of becoming a delta, and the properties of the delta must be determined, in turn, partly by the fact that it can become a proton. Generally, each one of the hadrons can communicate with all the others directly or indirectly through their numerous interactions. Therefore, the properties of each must be determined in part by those of the rest. In this sense each one of the hadrons is constituted of all the others.

This sense of kinship is strengthened by considering, not only the particles, but also the dynamics of their interactions. It is certainly significant that the particles of large mass interact with the strongest force and the greatest energy. Since the *differences* in mass values are certainly bought about by the interactions, it is not implausible to conclude that the particles themselves, whose mass values are not much larger than the differences between them, are of the same origin, that *interacting* and *existing* are but two manifestations of the same dynamic principles.

It has been proposed as an explicit program of particle research to treat the hadrons as creatures of their interactions. According to this program, which has been given the fanciful name of *bootstrap dynamics*, all the properties of particles and all of their interactions must come to be understood together as a self-determining whole in terms of basic dynamical laws.

These various methods for the theoretical analysis of particle phenomena are by no means independent. Group theory with its study of symmetries, S-matrix theory and Feynman diagrams, energy levels and bootstrap dynamics have all borrowed ideas from one another; and they have all used the principles of quantum mechanics and its elaboration in quantum field theory. It may well be that the development of a comprehensive theory of particles will incorporate concepts taken from all of these several approaches.

SPACE AND TIME

IN the *Principia* Newton gave definitions of space and time as he wished them to be understood for his discourse on mechanics. He defined "absolute" space and time thus:

> Absolute space, in its own nature and without regard to anything external, remains always similar and immovable.
> Absolute, true and mathematical time, of itself, and of its own nature, flows uniformly on without regard to anything external.

It is clear that he regarded "absolute" space and time as having objective existence, apart from any material bodies located in this space or experiencing changes with the flow of time.

He defined "relative space" as "some measure of absolute space, which our senses determine by its position with respect to other bodies," and "relative, apparent and common time" as "some sensible and external measure of duration estimated by the motions of bodies."

These definitions of space and time have been rightly criticized as vague and perhaps even meaningless. Newton was well aware of the inherent difficulties of attempting to define these most primitive entities, and he wisely asserted that he wished merely to counter certain popular misconceptions and to set up definitions adequate for scientific discourse. Further intellectual labors of philosophers and scientists have not succeeded in clarifying the ideas of space and time unequivocally. Three distinct aspects have been recognized: the metaphysical, the psychological and the scientific.

The philosophers of ancient Greece developed their concept of space and time primary to cope with metaphysical problems, such as those of existence, permanence and change. Their pronouncements were generally not enlightening. Aristotle defined time as "the measure of change with respect to before and after" and space as "the innermost limit of the containing body," which meant (to him) that space is a container surrounding objects. He had no difficulty in thinking of the whole of space as being within the outermost celestial sphere of the stars. Since this was thought to contain the whole universe, there was no space beyond it, an idea in marked contrast to Newton's absolute space, extending without bounds in all directions.

Philosophers following Newton were prone to discuss space and time in relation to their perception by the human mind, that is, as subjective entities. They were actually dealing with psychological problems at a time when psychology was still in a rudimentary state. Immanuel Kant (1724-1804) thought of space as the primary condition for the perception of spatial relations among real objects, as the "outer sense" for things external to the mind. He spoke of time, conversely, as the "inner sense" of contemplation, which could exist independently of the real world of space.

This pronounced dichotomy of space and time was mitigated later through improved knowledge of cognition. The American psychologist and philosopher William James (1842-1910) showed that, contrary to Kant, primitive spatial awareness is of material objects, of their sizes and shapes and relative positions and that abstract space is a more sophisticated conception, an intellectual ordering of spatial relations. The awareness of time is a similar ordering with regard to the *past* of memory, the *present* of immediate experience and the *future* of anticipation. Space and time thus serve similar functions of arranging experiences, respectively, in spatial and temporal sequences. This is analogous to their functions in science.

There is one notable difference between scientific time and the temporal sequence of which we are immediately aware. In all of the basic laws of motion the time t is treated as a fourth independent variable together with the three independent variables x, y and z, for length, width and depth in space. Just as these laws deal indiscriminately with left and right, they are equally valid for both directions of time; they recognize no distinction between past and future. Astronomers use the laws of celestial mechanics with equal facility to predict the occurrence of eclipses and to check astronomical observations of past centuries. Yet all human experience shows that time flows in one direction only.

To understand this disparity, it is necessary to digress briefly into a discussion of the laws of probability. One of these asserts that the probability of achieving a given state of affairs is proportional to the number of ways in which it may be realized. There are 26 ways, for example, in which a black card may be drawn by chance from a deck containing 26 red and 26 black ones. The state of having drawn a black card may be realized in 25 more ways by substituting each one of the remaining black cards for the one actually drawn. But the state of having drawn the ace of spades can be realized in one way only. This is why the drawing of any black card is 26 times more probable.

These considerations may be applied to the motions of balls on a billiard table, making as before (page 68) the simplifying assumption that all collisions are perfectly elastic so that the balls once set in motion continue to move forever. Consider first just 2 balls, one red and one white, whose random motions over the surface of the table are photographed by a moving-picture camera. If this film were shown on a screen in reverse, with time moving backward, the motion would appear to be quite possible and reasonable, an example of the indifference of the laws of motion to time reversal.

Assume, however, that there are 2000 balls, half of them red, the other half white, all those of one color being originally on either half of the table. Their random motion carries some balls of one color over among those of the other color; and this continues until the intermingling is complete, with equal numbers of balls of both colors on both halves of the table. If *this* process were photographed and the film projected on the screen with the end first, the time reversal would be immediately apparent, for the spectacle of 2000 mixed balls becoming unmixed spontaneously would be recognized as an impossible occurrence.

The difference between the case of 2 and 2000 balls may be understood in terms of probabilities. If the state of the 2 balls is described by noting on which half of the table they are located, it is evident that the state of having the white one on the left and the red one on the right can be realized in one way only. This is true as well for the reversed positions and for the states of having both balls on the right or on the left. This equality of probabilities is reflected in the intuitive expectation that the random motion might bring the balls into all of these states.

With 2000 balls the state of having the colors completely separated can still be realized in one way only. However, the state of having just one red among the white and one white among the red balls can be realized in as many ways as balls can be exchanged from one half of the table to the other while still maintaining this state. Each one of the 999 remaining white balls may be substituted, in turn, for the lone white one among the reds; and with each of these substitutions 999 red balls may be exchanged with the lone red one. Thus there are a thousand times a thousand, or a million ways, in which even this slight degree of mixing may be realized. With further intermingling the number of possible substitutions increases rapidly into the billions and trillions. Mixing is thus a transition from states of lesser to greater probability and has therefore a high probability of occurring; the reverse process is utterly improbable.

Another way of describing the matter is to note that the state of greater probability is also one of greater *disorganization*. The 2000 balls tend to go naturally and spontaneously from the well-organized state, with each color on its own half of the table, to the disorganized mixed state, but not the other way about.

It is important to note that the laws of mechanics do not assert that the unmixing of the 2000 balls is *impossible*, for the 2 balls, which mix and unmix freely in their random motions, are subject to these same laws. With large numbers the unmixing is simply so exceedingly improbable that it never occurs.

Processes of this sort, in complex systems with great numbers of individual parts, are called *irreversible*. Since all material things consist of myriads of molecules in incessant thermal motion, irreversible processes occur frequently in everyday affairs. The odor of flowers spreads through a room as the odor molecules mingle with those of the air. When a book pushed across a table slides to rest, the kinetic energy, which resided in the organized motion of all its molecules moving in the same direction, changes to the increased random motion of the molecules on the adjacent surfaces of the book and table top, warmed slightly by the friction which stopped the book. The energy is undiminished but it has gone from an organized to a disorganized state.

Irreversible processes are a part of everyone's common experience. They proceed incessantly and inexorably within our own bodies throughout our life span; and the immediate awareness of this constant trend toward the overwhelmingly probable gives our subjective sense of time its irreversible flow, from the past of memory to the anticipation of the future.

Particle phenomena are confined to such tiny regions of space and are so rapid and violent that each one occurs as if in isolation and is not sensibly affected by its surroundings. Thus, the number of entities involved in a particle interaction is small, and effects due to large numbers do not appear. In particular, since an interaction forward in time obeys all the relevant conservation laws, the time-reversed one must necessarily do so as well and should also be possible. But even here probability considerations may enter. When a proton and antiproton meet and create five pions, the reverse interaction requires the meeting of all five in the same point of space and at the same instant of time, a highly improbable event. Although the reverse process is possible in all cases, its realization may be so improbable that it is never observed.

There have been several alarums in experimental particle physics concerning time-inversion violation, but none have been actually confirmed. Yet after the shock occasioned by parity violation, physicists have become extremely cautious of making pronouncements about unexplored possibilities.

Certain conceptions of ancient Greek philosophers about the nature of space and time led them to an intriguing consequence concerning motion. While dealing with problems of the infinite divisibility of space and time, Aristotle and some of his contemporaries developed the idea that neither space nor time are truly smooth and continuous, that spatial extensions along a line is a sequence of distinct contiguous *points* and that similarly time is a sequence of distinct *instants*. A slowly moving object remains at one point for a large number of instants before moving on to the next while a faster one spends fewer instants at each point. This leads naturally to a *maximum possible* velocity in which the object moves from point to point at *each* successive instant. Since this outcome was held to be patently absurd, the idea of discrete natural units of space and time was abandoned. Today it is well known that there *is* a maximum possible velocity.

Modern conceptions of particle physics and quantum mechanics have yielded conjectures remarkably similar to those of Aristotle. In particular, it was suggested as early as 1928 that, as a consequence of the discrete value of Planck's constant, there should be a natural "quantum" of time, the *chronon*. Various numerical estimates yielded values for the chronon of about 10^{-24} second, comparable to the lifetimes of the virtual pions and vacuum particles shown in the Feynman diagrams on page 202. A stationary particle may be thought of as going into, and out of, existence while remaining at the same place; one which is moving is annihilated at one point and created anew at the next. Since the highest possible velocity is that of light, the intervals between the discrete points of space (assuming time intervals of about 10^{-24} second) cannot exceed 10^{-14} centimeters, a value which might be taken as the ultimate unit or "quantum" of space. Since a particle does not "move through space" between adjacent points, there need be nothing, not even space between them.

Whenever experimental procedures reach a degree of nicety comparable to the ultimate units of nature, phenomena may undergo profound changes. Wind tunnel experiments performed with a small model airplane simulate the full scale effects if the reduction in size is tenfold or a hundredfold. But if the size were decreased by a factor of a hundred million, making it comparable to the diameter of air molecules, the model would behave like a real airplane pelted on all sides by cannon balls; it would be buffeted about in a violent and unpredictable manner. Similarly, a very minute electric current acts not as a smooth flow of "electric fluid" but as a randomly fluctuating passage of individual electrons; and a ray of light of very low intensity is found, by means of a sufficiently sensitive detector, to be a stream of discrete photons.

It may well be true that particle research is penetrating into an order of minuteness in which the ultimate "graininess" of space and time is becoming apparent. This could be the underlying cause for some of the peculiar difficulties of particle theories.

SUMMARY OF PARTICLE THEORIES

IT is evident that the progress of basic theoretical research in physics over the past several decades has greatly broadened the horizons of our knowledge concerning the nature of the physical universe. It has shown that the stuff of the universe is surprisingly multifarious, that it comprises a large number of particles having complex properties and equally complex ways of interacting with one another. The existence of virtual particles as the quanta of force fields has mitigated the action-at-a-distance difficulty and has restored a measure of reality to the field concept. Even "empty" space is replete with these evanescent particles in a curious accord with the medieval dictum that "nature abhors a vacuum."

New conservation laws have been recognized which, together with the older familiar ones, are capable of accounting for many of the characteristics of particle phenomena. Whereas familiar experience implies that phenomena are reversible in space but not in time, the study of particles has shown that just the opposite is true, that reversal in time is always possible while some kinds of happenings are impossible when reversed in space, as in a mirror. Spatial symmetry is restored in the world of antiparticles, which may be thought of, perhaps only in an abstract sense, as one in which all events proceed backward in time.

Theoretical studies have revealed various kinds of organizations and relations among particles both in their properties and in their interactions, and have shown that these two aspects are not unrelated, that existing and interacting may be two manifestations of the same basic causes functioning within one consistent, self-determining whole.

Yet with all these new insights more questions have been raised than answered, and the unanswered ones are of a most fundamental kind. There is the question of mass values. Certain relations have been discovered among some of these values, but not one of them has been calculated in terms of a basic theory. Similar regularities have been found in the distribution of charge values, in singlets, doublets, triplets and quadruplets, but here again a basic explanation is lacking. While it seems that most particles have an internal structure, various ways of describing it lead to discordant concepts.

Conservation laws determine what interactions can and cannot occur, but the relative probabilities for the various possible ones have not been explained, nor is it known why the strongest force is restrained by the greatest number of conservation laws. These matters will probably not be understood until more is known about the nature of particle forces. The new conservation laws, of strangeness, baryon number, and so forth, are still obscure; and the symmetries to which they might be related are unknown.

The recent history of physics suggests that progress toward the solution of these problems may require the discovery of a radically new idea such as Planck's quantization of energy, in 1900, and de Broglie's matter waves, in 1924. There is, however, no assurance that this seminal idea will be any more than a successful schema for making all the necessary mathematical calculations. Physicists have come a long way from the attitudes of the eighteenth century, when a scientific theory was not considered adequate unless it was given in terms of a mechanical contrivance whose every linkage and action could be pictured in detail. The idea of "really understanding nature" in this naïve sense has become untenable, particularly in physics, the science which deals with the most primitive and fundamental aspects of nature. The goal of contemporary physics is more modest: a consistent program of mathematical procedures which accepts certain well-established characteristics of physical systems and generates numerical values in close agreement with experimental observations on those systems. This kind of a description of physical phenomena is accepted as satisfactory today, though perhaps with some private reservations, by the majority of physicists.

Experimental particle research is exceedingly costly and is certain to become more so in the future. Each advance in experimental capabilities will no doubt produce new knowledge to make that advance obsolete in the forefront of basic research. Furthermore, experimental particle research, as it has been pursued well into the second half of this century, has been almost completely devoid of any useful applications to technology and industry.

The history of science shows that *pure* research, having no other purpose than the advancement of knowledge, has often paid off more handsomely in practical results than *applied* research, aimed directly at attaining such results. In fact, an important scientific discovery is usually characterized by appearing unprepossessing and of no practical value. A new supersonic airplane or a moon rocket does not fulfill these requirements. These are technological achievements rather than scientific discoveries. A true scientific discovery contains an element of surprise; it reveals something quite unsuspected about the workings of the universe. Sooner or later this revelation may lead to useful application, perhaps far removed from the discover's field of interest. Thus, X-rays have proven of tremendous value in surgery, but Wilhelm Roentgen (1845-1923) would never have discovered them if he had set out to make improvements in surgical techniques. Alleviating the housewife's burdens was probably furthest from the mind of Michael Faraday (1791-1867) while observing the interaction of electric currents and magnets at the Royal Institution. Yet the force which caused a bit of wire near a magnet in his laboratory to jerk slightly when a current was passed through it is the same force which runs the electric motors in many household appliances, and much of the machinery of industry and transportation as well.

The primary justification of all basic research lies in the increase of human knowledge, and particle research certainly ranks high in this regard. The investigation of the stuff of which the universe is constituted has been recognized in all ages as a worthy object of intellectual quest.

Looking upward and outward into the heavens has always held a special appeal for the inquiring human spirit. But much of the knowledge achieved in our century about what happens in the astronomical reaches of space has been learned by peering downward and inward into the innermost recesses of matter. Particle research has given new objectives and new impetus to cosmology, the science whose province is the universe as whole, its origins, its development and its ultimate destiny.

There is in the universe a hierarchy of forces inversely proportional in strength to the size of the regions in which they manifest their action. There is first of all the strongest force acting in the smallest of stable structures, the atomic nuclei. Its great strength assures the integrity of its domain against all ordinary attacks. Yet this force is so closely limited in range that it cannot interfere with the action of the electromagnetic force, next in order of decreasing strength. Alone of all known forces this one shows a dual action of repulsion as well as attraction, and it is the balance of the two which gives each kind of atom its unique size and shape and mode of interaction with the surrounding matter and energy. This is the force which determines the physical and chemical properties of all familiar material things. It is, furthermore, of most immediate concern to ourselves, for it determines the minutest details of our bodily structures, and it mediates all the functions of our muscles, nerves and brain. Leaving aside the weak force, with its special and limited role in the scheme of things, there is finally the gravitational force, inherently so weak that it is manifest only with objects of great size, yet capable of reaching out for enormous distances to dominate the vast universe of planets and stars and galaxies.

Isaac Newton, whose work stands at the beginning of our story, spanned the scientific progress of three centuries with these prophetic words:

> Now the smallest particles of matter may cohere by the strongest attractions, and compose bigger particles of lesser virtue; and many of these may cohere and compose bigger particles whose virtue is still weaker, and so for divers successions. . . . There are therefore agents in nature able to make the particles of bodies stick together by very strong attractions. And it is the business of experimental philosophy to find them out.

15

Philosophical Implications of the New Physics

SCIENCE is more than an accumulation of facts about the physical universe. Facts are the building stones out of which the edifice of science is constructed. But the construction requires study, discrimination, planning, judgment and imagination. It involves both the external world of observed things and the inner world of the observer's mind; and what is built is determined in no small measure by the nature, the capabilities and limitations of the human intellect. It is not surprising that thoughtful men of science have, from the very beginnings of their discipline, been impelled to give consideration to the philosophical problems of the mind and its relation to the external world.

As a private individual and a member of society, a scientist may well have a concern for *esthetics*, the philosophy of beauty and its expression in art, and for *ethics,* the philosophy of conduct, of right and wrong, of morals and responsibilities. But in his professional activities his primary philosophical interest is necessarily in *metaphysics,* for metaphysics is concerned with the way things are rather than with the way things ought to be.

Metaphysics falls into two divisions, corresponding roughly to the dichotomy of the inner world of senses, thoughts, and ideas and the external "real world" of material objects. *Epistemology* is concerned with the former. It deals with questions such as: What is knowledge? How is it acquired? How can it be affirmed? In what ways might it be limited? *Ontology* is the philosophy of the world of things. It inquires into the nature of reality, in what way and why things are considered to be real. Man and his mind are a part of reality. Therefore, ontology and epistemology are not distinctly separable.

Metaphysics and science thus relate to the same areas of knowledge. But whereas science formulates statements about the nature of the physical world, metaphysics is concerned with the nature of these statements, with their meaning, their relevance to cognition, their logical implications and internal consistency.

The scientific activity of merely observing nature might well be done without recourse to philosophy. But observations *per se* are meaningless unless they are related to a plan or purpose, such as the proof or disproof of a tentative hypothesis. A biologist might spend a lifetime measuring and recording the height of every tree he sees; but his data would have value only if they were related to some pertinent factors of growth, such as soil conditions or mean annual temperature. Here is involved the whole complex of problems about the ways in which observations are to be evaluated, what criteria should be used in deriving generalizations, how these may be affected by objective evidence, scientific traditions and individual predilections, and many others in the area of epistemology—problems which have been studied and clarified through centuries of philosophical labors.

At an even more fundamental level the scientist must realize that all he can be directly aware of are sense impressions, such as those of color, form and spatial relations, of sounds and tastes and odors, of tactile sensations of hot and cold, texture and hardness, all of which exist somewhere within himself, in his immediate consciousness. Whether or in what way these are occasioned by an objective reality external to himself is again a philosophical problem, a problem of ontology to which philosophers have devoted much disciplined thought.

In particular, where the scientist has advanced, as has the theoretical physicist in recent times, far toward the limits of the comprehensible, he must understand that his activities involve his inner self as well as the world external to him. He must give thought to the workings of his own mind and distinguish carefully between his sense impressions and his interpretations of them; whether, for example, a significant order in events lies in the observed objects or is placed there by his subjective rationalizing. With a background of relevant knowledge of metaphysics the scientist is better-equipped to cope with problems, such as the relation of the observer to the observed events, which are coming into prominence in the forefront of physics research.

The following discussions lie in the realm of philosophy; and as will be quickly apparent, their tenor is rather different from that of the previous considerations of scientific matters. Philosophers necessarily have a broader outlook than scientists of any one discipline. They speculate more boldly and are apt to depart more widely from the solid ground of observed facts. Philosophical propositions are arrived at by logical reasoning and are, in principle, not debatable. But the premises from which the logical arguments proceed are often personal opinions and convictions. Thus, philosophers of various eras and different countries have frequently come to highly disparate conclusions, and many philosophical questions have been taken up repeatedly and debated at great length over periods of centuries.

In ancient times and throughout the Middle Ages science or *natural philosophy* was not set apart from other branches of philosophy. But after the time of Francis Bacon and René Descartes science developed increasingly into a separate discipline. At the same time philosophers became more scientific and more individualistic in their outlook. They observed the state of the world about them and formulated their own conclusions concerning the condition of man and of nature. Epochal advances in science, as those achieved by Newton in physics, Darwin in biology and Freud and his successors in psychology, exerted a strong influence on the philosophers of their times.

The following discussions are concerned largely with the ways in which certain philosophical concepts have been affected by the new developments of atomic physics in this century. Most of the questions to be considered had already been put in good order by the work of competent professional philosophers, a fortunate circumstance which is used to good advantage here.

But in the decades following immediately upon the stirring discoveries of the new physics, these were given enthusiastic but ofttimes misguided interpretations, some of which are still evident in popular writings about science. The attempt to set these matters aright is by no means a tilting at windmills.

To establish a proper background for the clarification of several difficult concepts which need to be considered, their history and development will be traced from their origins. These surveys will be held to a reasonable scope through a somewhat arbitrary but hopefully representative selection of pertinent discussions and excerpts from the writings of philosophers and scientists.

SPIRITUAL VALUES IN THE OLD AND THE NEW PHYSICS

THROUGHOUT the eighteenth and nineteenth centuries, during which classical science scored its greatest triumphs, it was not without critics and detractors. They were by no means consistent, some damning science for characteristics which others found praiseworthy. Yet the tenor was that science is cold and materialistic and hostile to, or at best neglectful of, man's higher aspirations, his sensibilities in esthetics and ethics and his religious feelings of awe and reverence. Even today some speak boastfully of their lack of interest in science as if this were *ipso facto* an indication of their concern for the finer things in life.

Classical science is materialistic in that, up toward the end of the nineteenth century, it strove to explain all phenomena in terms of mechanisms constructed of material parts. Yet in Newton's day this did not prevent scientists from maintaining an active interest in philosophy and theology, with art, architecture, belles-lettres and statesmanship added for good measure.

Newton himself devoted more time and thought to theological than to scientific matters, and even in his scientific writings God's presence in the world was not forgotten. When he could not find in the physical world a universal fixed frame of reference for all motions, he concluded that it must reside "in the sensorium of God." In contemplating the dynamics of the solar system, he was aware that the slight gravitational forces acting between the planets must produce small perturbations in their elliptical orbits which over long periods of time could cause instability and collapse. He concluded that God could not permit the universe to go on by itself but would have to intervene at intervals of many millennia to set things aright.

René Descartes went much further in frankly basing all scientific work on metaphysical concepts. He made the existence of God a precondition for all of science. Seeing that there was the logical possibility that all his beliefs were false, he sought for a solid base upon which he might build an edifice of truth. He found that the one fact he could be sure of was his own existence as a thinking being: "Cogito, ergo sum." Since he recognized that he was finite and imperfect, he reasoned that this sense of his limitations implied that there had to be an infinite and perfect Being against whom he measured himself. This Being had to exist because He would not be perfect if he lacked the essential attribute of existence. Thus Descartes proved the existence of God in a manner which had for him all the force and rigor of a mathematical demonstration. To him this result was more important for science than for theology, for it afforded the possibility of solving the central problem of ontology, the problem of the nature of reality. Since all our awareness and knowledge of a world external to our minds come to us through sense impressions, the question arises whether there exists anything other than just these, whether objective reality is an illusion. To this question Descartes answered that God, being perfect, cannot be a deceiver, that he would not make men believe in the existence of a material universe if it were not real.

Thus, the reality of the realm in which science must work is assured because of the existence of God. In his *Meditations*, published in 1641, Descartes writes "the certainty of all things depends on it [the existence of God] so absolutely that without this knowledge it is impossible to know anything perfectly."

Descartes was an eminent example of the universal savant. His philosophical works had such a wide and lasting influence that he has been called "the father of modern philosophy." In science his ideas about biology, physics and astronomy were widely discussed. But above all his fame rests on his achievements in mathematics, on his development of *analytical geometry*, one of the most fertile conceptions in all of mathematics.

As the sciences proliferated greatly in the eighteenth century, both in range and complexity, the universal savant became an impossibility. It was necessary to opt either for science or for philosophy, and this trend toward specialization became increasingly evident. Secular philosophy and theology developed into separate disciplines, as did the physical and biological sciences. In time these were again splintered into ever-narrower specialties. After the transition into the nineteenth century references to theology and ethics, and all nonscientific matters, became rarer and gradually disappeared entirely from scientific writings. This does not imply, however, that scientists relegate esthetics, ethics and religion to an inferior status compared to their professional concerns.

In one important respect classical science did clash with theology, or more correctly, with theological pronouncements concerning matters in the realm of the sciences. The theological doctrines of all religions include statements relating to cosmology, astronomy, geology, anthropology, biology and the various other branches of science. The accounts given by some religions are highly fanciful and bizarre. The early Hebraic-Christian conception of the universe is by contrast a remarkably logical structure based upon careful observations.

If a thoughtful, primitive observer, ignorant of all the knowledge which has been accumulated over twenty centuries of scientific progress, were to stand on a high hill looking out upon the earth and the heavens to where they seem to meet at the far horizon, he might well come to conclusions such as these:

> The earth is by all odds the largest object in the entire universe.
> The earth is flat. It is stationary and limited in extent.
> The sun is incomparably smaller than the earth, yet much larger and brighter than any of the stars.
> The sun and moon move across the heavens above the stationary earth.
> The change from bright day to dark night occurs at the same time throughout the entire universe.

Furthermore, if his knowledge about the nature of things were based on the observations made over his own life span, plus the verbally transmitted history going back a few generations, he would believe that all of his environment has remained essentially unchanged with the passage of time; that the earth with its hills and valleys, its rivers and seas, and the sun and the stars in the sky above have always been and will always remain the same; that the living creatures he is familiar with have existed forever and that the cycles of the seasons, of growth and decay, of birth, life and death from an ever-recurring, changeless pattern of events.

Clearly, the story of Genesis is in harmony with these concepts. Yet none of them agrees with what is now known about the nature of the universe. In common parlance they would be said to be wrong. "Wrong" is, however, hardly the word to characterize a statement based on careful, though limited observations. Here it is instructive to consider the simple example of plumb bobs hung side by side on cords as shown in Figure 36. Careful measurements made on two of them hanging close together would show that they are parallel, that is, the cords are the same distance d apart at the top and bottom. If a series of bobs were hung at stations along a meridian from the North Pole to the Equator, observations would show that the cords are all directed toward the center of the earth, that they are not parallel. Thus, observations of nearby plumb bobs lead to the law: "Plumb bobs hang with their cords parallel," while more extended observations show: "Plumb bobs hang with their cords pointing toward the center of the earth." The first law is not "wrong" (masons do not hesitate to use it when putting up a wall); rather, it is valid only over a restricted range.

Similarly, classical mechanics has been found to give an adequate account of physical phenomena only over a limited range of velocity and of size. Relativistic mechanics holds over a wider range of velocities up to that of light, and quantum mechanics is valid for systems of atomic size. Both Einstein and de Broglie would use the laws of classical mechanics in the range in which they have been thoroughly verified by experiment.

The conclusions of our primitive observer have all been shown to be inadequate through the broader knowledge of modern science, knowledge obtained in good part through work of the paleontologists and the use of the telescope, the microscope, the photographic camera, the spectrograph, the radio and the many other devices which science has developed to extend the range, the

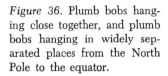

Figure 36. Plumb bobs hanging close together, and plumb bobs hanging in widely separated places from the North Pole to the equator.

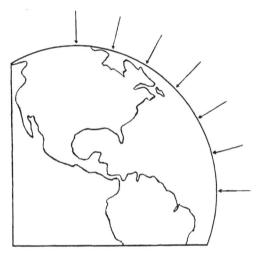

sensitivity and the precision of the human powers of observation. We now know that the earth is but an infinitesimal speck in a huge universe, that it is exceeded greatly in size and mass by the sun, which is by no means the largest and brightest star. We know that the earth is a sphere over whose surface one could wander endlessly without ever toppling off the edge. We know that, as it turns about its axis, one region after another comes into the light of the sun so that it may be night in London and day in Tokyo at the same time, and that indeed the succession of night and day has no meaning for the universe as a whole, where it is constantly bright day near any one of the stars and constant night in the far reaches of space between them. We know also that the vast majority of animate creatures have come into being, have multi-

plied and prospered for a time only to meet insurmountable difficulties and pass into extinction.

Change, not permanence, is the essence of all things. Between
two ticks of the clock our sun radiates a million tons of its substance into the depths of space, never to return. With the passage
of time the eternal hills wear away, and the eternal stars have a
birth, a span of life and a descent into darkness.

If a new religion were to arise today with a theology which
includes pronouncements about the nature of the physical universe, framed as thoughtfully as those of the Old Testament, it
would contain references to RNA molecules, de Broglie matter
waves and to many other findings of contemporary science. But
if the founders of this new religion were to assert that these statements are eternal verities, they would be at loggerheads a hundred years later with the scientists who would have advanced to
newer knowledge and more broadly valid interpretations.

A controversy between the two could arise because both would
be contending in the realm of science, and they could also fall
into arguments if the scientists were to make arbitrary assertions
about matters of theology. But a clash between pronouncements
about theology and about science, each pertaining to its proper
concern and purposes, is logically impossible. Whether the universe evolved from an intelligent plan or from the blind workings
of material forces, whether there exists a Being who has a concern
for the human condition, whether man has an immortal soul,
whether feelings of love, awe and reverence have a basis in reality
—these are questions which may be affirmed or negated quite
apart from any evidence which science might advance concerning
the nature of the physical universe.

Still, clashes did occur, often carried on with great vigor and
acrimony. The contestants did not always hold to the niceties of
logic, and statements about God and the nature of the universe
were not always assigned to their proper categories of science and
of religion. Thus, the concept of a three-layer universe, with
heaven above, hell below and the earth in between, depicted
dramatically in numerous religious paintings, is a cosmology; and
it could therefore be demolished handily by contrary scientific
evidence.

Some of the most violent controversies occurred in biology.

Here, in particular, many warnings were sounded against the encroachment of materialistic science into the realm of the spirit. Thomas Henry Huxley (1825-95), English biologist and enthusiastic apologist for scientific objectivity, commented on this subject in an address, "On the Physical Basis of Life," delivered at Edinburgh in 1868:

> As surely as any future grows out of the past and present, so will the physiology of the future gradually extend the realm of matter and law until it is co-extensive with knowledge, with feeling and action. The consciousness of this great truth weighs like a nightmare, I believe, upon the best minds of these days. They watch what they believe to be the progress of materialism, in such fear and powerless anger as a savage feels, when, during an eclipse, the great shadow creeps over the face of the sun. The advancing tide of matter threatens to drown their souls; the tightening grasp of law impedes their freedom; they are alarmed lest man's moral nature be debased by the increase of his wisdom.

Huxley was referring to the long struggle between the proponents of *vitalism* and *mechanism*, in which the question at issue is whether the phenomena occurring in living organisms are governed by the same laws which hold for animate matter, or whether these phenomena are mediated by a special "vital principle." According to the latter view, chemical compounds were divided into *inorganic* and *organic*, the latter requiring the special vital principle within living organisms for their synthesis. This was maintained stoutly by the Swedish chemist Jöns Jakob Berzelius (1779-1848). But it was challenged by one of his own students, Friedrich Wöhler (1800-82), who in 1828 synthesized urea, a typical organic compound; and it was thoroughly disproved when, some two decades later, the French chemist Pierre Berthelot (1827-1907) produced dozens of organic compounds in his laboratory.

The idea of mechanism was already old in Huxley's time. More than two centuries earlier Descartes had asserted that all animals are machines. But this did not arouse "fear and powerless anger" because man was excluded as a matter of course from any relation to all other forms of life. Mechanistic ideas about living things became a heated issue only after Darwin had brought them into collision with human vanity. Curiously, those who were most vociferous in rejecting all kinship with the rest of living organisms were also most insistent upon keeping the latter uncontaminated by relations to the animate world.

From the vantage ground afforded by a century of progress in biology, it appears somewhat surprising that the synthesis of complex molecules in the test tube should have aroused fear and anger, or that the validity of physical laws for the mechanical, thermodynamic, electrical, optical and chemical processes in the human body should be thought debasing to man's nature.

Classical science has been criticized, however, more recently and on much broader terms. The English astronomer and mathematician Sir James Jeans (1877-1946) in his book *The Mysterious Universe*, published in 1930, describes the universe of classical science in these dramatic words:

> Standing on our microscopic fragment of a grain of sand, we attempt to discover the nature and the purpose of the universe which surrounds our home in space and time. Our first impression is something akin to terror. We find the universe terrifying because of its vast meaningless distances, terrifying because of its inconceivably long vistas of time which dwarf human history to the twinkling of an eye, terrifying because of our extreme loneliness, and because of the material insignificance of our home in space—a millionth part of a grain of sand out of all the sea sands of the world. But above all else, we find the universe terrifying because it appears to be indifferent to life like our own; emotion, ambition and achievement, art and religion, all seem equally foreign to its plan. Perhaps we ought to say it appears to be actively hostile to life like our own. . . .
>
> Into such a universe we have stumbled, if not exactly by mistake, at least as the result of what may properly be described as an accident.

This, Jeans avers, is the conception which classical science has given of man's place in the universe; and from it Jeans concludes:

> Viewed from a strictly material standpoint, the utter insignificance of life would seem to go far toward dispelling any idea that it forms a special interest of the Great Architect of the universe.

While the dramatic style of these passages may have an immediate emotional appeal, his argument is found upon sober reflection to be largely gratuitous. If it were indeed true that living, sentient beings exist nowhere else in the universe but on our tiny earth, a supposition which had better scientific warrant at the time of Jeans's writings than it has today, the grandiose setting which the "Great Architect" had provided for these beings could as well be interpreted as a proof for his special concern for them. Jeans is, of course, correct in saying that living things, as reckoned by the ton, are exceedingly rare. But scarcity does not necessarily detract from value; the rarest flowers are treasured most highly. Further, if relative size is to be advanced as a measure of worth, man's huge bulk compared to that of the atom could be an impressive argument for his importance in the scheme of things.

Blaise Pascal, the French scientist and philosopher, saw no reason for feelings of inferiority in man's physical stature. In his *Pensées* (published posthumously, in 1670) he wrote:

> Man is but a reed, the most feeble thing in nature. But he is a thinking reed. . . . If the universe were to crush him, man would still be nobler than that which destroyed him, because he knows that he dies and the advantage the universe has over him. Of this the universe knows nothing.

Jeans is well worth quoting, however, for he is a widely read writer of popular books on science, and his ideas are typical of his time, of the early years of our century.

In Newton's day the prevailing attitude of scientists toward their discipline was far different. The founders of classical science, the men who brought the Royal Society into prominence, saw a deeply religious purpose in their studies: the glorification of God through the revelation of his works.

Robert Boyle, who refused the nomination to the presidency of the Royal Society because the oath of office conflicted with his religious principles, expressed the prevailing concept of scientific investigations thus:

> God loving, as he deserves, to be honored in all our faculties, and consequently to be glorified and acknowledged by the acts of reason as well as those of faith, there must surely be a great

disparity betwixt the general, confused and lazy idea we commonly have of His power and wisdom and the distinct, rational notions of these attributes which are formed by an attentive inspection of those creatures in which they are most legible, and which were made chiefly for that very end.

This idea, that God created the world and gave man the intelligence to study it and thereby reveal His "power and wisdom," gave scientific labors a religious purpose and accounted for the close union of scientific and spiritual endeavors both in the lives of individuals and in the general attitudes of these times.

How in the course of three centuries this close relation gave way to separation and even antagonism is a long and complex story. Some of its elements, increased specialization and the conflict between theological and scientific views about the physical universe, have already been touched upon. More potent was the general conviction, prevalent among scientists and laymen alike, that science in the Victorian Age had become a thoroughly explored realm, no longer offering worthy challenges to the human intellect, and that in its very maturity science had nothing to offer concerning the most basic and poignant aspects of the human condition. Science came to be looked upon as cold and wholly objective, a structure of rigid deterministic laws, hostile to freedom of the human spirit. Classical science had grown old; and the brave expectations of its founders, that it could bring a clear understanding of God and of all his works, had not come to pass.

On the other hand, the tremendous success and prestige of science in the material realm, the increased understanding and control of natural forces, the lessening of superstitious fears, and above all the bold spirit of uninhibited doubting and testing served but to exacerbate the breach between theology and science.

The concern over this breach is clearly revealed in the works of those nineteenth-century writers who deplored it and strove to heal it. The Scottish biologist and philosopher John Scott Haldane (1860-1936), named by his contemporaries "the chief liaison officer between science and philosophy," points out in his essay "The Sciences and Religion" that the relations of science and religion have been the concern of all great philosophers since Descartes. But, he says, scientists have remained largely ignorant of this problem; and "even now we find scientific writers taking actual pride in their ignorance of philosophy. They are in a similar position to that of the schoolmen who despised experimental science." Theologians, on the other hand, have conceded to science only a narrow materialistic view of nature, which excludes spiritual values. Thus, the two groups have fallen into an unnecessary conflict. Haldane stated:

> "The present widespread belief that religion will die out as science advances is nothing but evidence of intellectual blindness: Existing churches will decay if they do not amend their creeds; but religion will no more die out than science will or philosophy will.

All of this has to modern ears a quaint and faintly archaic sound, as of a controversy long since laid to rest. Of greater present interest are the ways in which the new physics of the twentieth century has been represented as reestablishing subjective spiritual values in the fundamental principles of science. In particular, it will be of interest to discuss the ideas expressed by scientists and philosophers, and also by various popular authors, concerning the metaphysical and theological implications of the changes wrought by de Broglie, Heisenberg, Bohr, Schrödinger, Born and others in the eventful decades following the birth of quantum mechanics.

These new interpretations are well exemplified by the later passages of Sir James Jeans in *The Mysterious Universe*. Tracing in broad outlines the progress of ideas about nature, he maintains that the earliest conceptions were essentially anthropomorphic. "Confronted with a natural world which was to all appearances as capricious as himself, man's first impulse was to create nature in his own image. . . ." More intelligent and systematic observations showed, however, that many events proceed in an orderly manner. This led to the idea of natural laws, which culminated in the seventeenth century in the mechanistic, deterministic philosophy of classical science. Jeans quotes Hermann von Helmholtz (1821-94), the German physiologist and physicist, who declared that "the final aim of all natural science is to resolve itself into mechanics."

Then come Planck's quantized oscillators, Einstein's photons and Bohr's atoms with their quantum jumps for which, says Jeans, science "cannot predict with certainty which state will follow which; this is a matter which lies on the knees of the gods." Radioactivity is given as a further example of indeterminism in that "the individual radium atom does not die because it has lived its life, but rather because in some way fate knocks at the door." As a result of such observations, it appears that "what we are finding, in a whole torrent of surprising new knowledge, is that the way which explains them more closely, more fully and more naturally than any other is the mathematical way, the explanation in terms of mathematical concepts."

All of this development Jeans sums up as follows:

> Our remote ancestors tried to interpret nature in terms of anthropomorphic concepts of their own creation and failed. The efforts of our nearer ancestors to interpret nature on engineering lines proved equally inadequate. . . . On the other hand, our efforts to interpret nature in terms of the concepts of pure mathematics have, so far, proved brilliantly successful. It would now seem to be beyond dispute that in some way nature is more closely allied to the concepts of pure mathematics than to those of biology or of engineering. . . . We have already considered with disfavor the possibility of the universe having been planned by a biologist or an engineer; from the intrinsic evidence of his creation, the Great Architect of the Universe now begins to appear as a pure mathematician.

In this latest outcome Jeans sees a close kinship between man and the physical universe, and "it can hardly be disputed that nature and our conscious mathematical minds work according to the same laws." Somewhat cautiously he adds that "the universe can best be pictured, although still very imperfectly and inadequately, as consisting of pure thought, the thought of what, for want of a wider word, we must describe as a mathematical thinker."

These developments, Jeans concludes, alter radically our conception of the universe and of man's relation to it:

> Thirty years ago, we thought, or assumed, that we were heading toward an ultimate reality of a mechanical kind. . . . Into this wholly mechanical world . . . life had stumbled by accident.

Today there is a wide measure of agreement, which on the physical side of science approaches almost to unanimity, that the stream of knowledge is heading toward a non-mechanical reality; the universe begins to look more like a great thought than a great machine. Mind no longer appears as an accidental intruder into the realm of matter. . . .

The new knowledge compels us to revise our hasty first impression that we had stumbled into a universe which either did not concern itself with life or was actively hostile to life. The old dualism of mind and matter, which was mainly responsible for the supposed hostility, seems likely to disappear . . . through substantial matter resolving itself into a creation and manifestation of mind.

Finally, Jeans cautions that the "river of knowledge is a winding one" and that "No scientist who has lived through the last thirty years is likely to be too dogmatic either as to the future course of the stream or as to the direction in which it lies. . . ." He concludes "that our main contention can hardly be that the science of today has a pronouncement to make; perhaps it ought rather to be that science should leave off making pronouncements. . . ."

There is some justification for characterizing the new physics as mathematical rather than mechanical. On the other hand, classical physics is not the monolithic mechanical structure which Jeans seems to make of it. There are important areas of classical physics which are predominantly mathematical. For example, classical thermodynamics, which deals with the relations of heat and mechanical energy, is a body of mathematical laws with no essential relations to mechanical concepts. Although Maxwell used a mechanical ether model while developing his electromagnetic theory, the completed structure is essentially mathematical; the "physical reality" to which the equations relate is a vague, non-material concept of electromagnetic fields. Even Newton was accused of having built in his laws of mechanics a purely mathematical structure without physical content.

Further, there arose in the latter half of the nineteenth century the philosophy of *energetics*, advocated by such influential men as the chemist Friedrich Ostwald (1853-1932) and the physicist Ernst Mach (1838-1916) in Germany and the physicist and engineer William Rankine (1820-72) in England. They repudiated a description of the universe in terms of mechanical particles and asserted that energy relations must form the underlying description for all physical phenomena, an assertion which the French philosopher and scientist Pierre Duhem (1861-1916) thought to interpret as evidence for the reality of mental and spiritual energy.

It is nevertheless true that the new physics has turned away in a most thorough fashion from mechanical models and toward mathematical description, and that this may well be a transition from which there is no return. Physics has proceeded into realms so far removed from ordinary experience that its theories may never again be couched in other than abstract mathematical symbols.

Paul Dirac, the English physicist who contributed importantly to the building of the new physics, expresses this clearly:

> In the case of atomic phenomena no picture can be expected to exist in the usual sense of the word "picture," by which is meant a model functioning essentially on classical lines. One may extend the meaning of the word "picture" to include any *way of looking at the fundamental laws which make their self consistency obvious*. With this extension one may acquire a picture of atomic phenomena by becoming familiar with the laws of quantum theory.

Although Jeans's dramatic style does produce specific statements which may be challenged, he gives on the whole a good account of how the new physics has altered the scientific conception of the universe. He avoids dogmatism and does not go beyond the proper bounds of science. While he stresses that "the universe shows evidence of a designing or controlling power that has something in common with our individual minds," he specifically excludes such mental activities as "emotion, morality, or aesthetic appreciation." Other writers, however, have not hesitated to include not only ethical, esthetic and spiritual values but also mystical religious experience among the advantages which the new quantum physics has conferred upon the material universe. Indeed, classical physics is pictured as an evil dragon so that de Broglie, Heisenberg, *et alii*, may be cast in the role of knights in shining armor, riding to the rescue of the beleaguered human spirit.

This trend is evident in the writings of the English author J. W. N. Sullivan, a popular interpreter of science and philosophy. In *The Limitations of Science* (1933) he contends that in classical physics:

> . . . only those elements which acquaint us with the quantitative aspects of material phenomena are concerned with the real world. None of the other elements of our experience, our perception of color, etc., our response to beauty, our sense of mystic communion with God, has an objective counterpart.

"Our religious impulses," he continues, "cannot be satisfied with anything less than a belief that life has a transcendental significance. And it is precisely this belief that the old philosophy of science made impossible."

In recent times, however, says Sullivan: "Science has become self-conscious and comparatively humble." In particular, the new science does not require that we "know the nature of the entities we discuss, but only their mathematical structure." This, he believes, has important consequences:

> The fact that science is confined to a knowledge of structure is obviously of great "humanistic" importance. For it means that the problem of the nature of reality is not prejudiced. We are no longer required to believe that our response to beauty, our mystic's sense of communion with God have no objective counterpart. It is perfectly possible that they are, what they have so often been taken to be, clues to the nature of reality. . . . Our religious aspirations, our perceptions of beauty, may not be the essentially illusory phenomena they were supposed to be. In this new scientific universe even mystics have a right to exist.

Sullivan is correct in asserting that the new quantum mechanics emphasizes "mathematical structure." He might compare the "prejudged reality" of the Bohr atom model's neatly depicted electron orbits to the Schrödinger wave patterns in abstract multidimensional space. There is, however, the question of how much reality should be attributed to the Bohr model. While some authors of popular science writings have represented it as having actual physical existence, Bohr himself thought of it as no more than a tentative approximation to reality, whatever that might be.

The further sweeping indictment of classical physics as depriving life of "transcendental significance" and denying the mystics' "right to exist" is more than a little curious.

There is nothing unique to the new physics in this juxtaposition of "physical reality" and "mathematical structure." In the mid-nineteenth century Maxwell's equations of the electromagnetic field effectively abolished the "prejudged reality" of the ether, though it took half a century for this to be generally understood. On the other hand, classical thermodynamics, a mathematical structure with no prejudged reality (in it gases were characterized vaguely as "tenuous continua"), was later fleshed out by the detailed reality in the kinetic theory of gases pictured as swarms of randomly moving molecules, a reality which was widely accepted as a signal advance in scientific comprehension. Furthermore, at the very beginning of classical science Newton opposed his mathematical description of the solar system to the prejudged reality of Descartes's ether vortices, a development with which contemporary philosophers were less than enchanted.

Physicists are moreover not of one mind concerning the nature of "reality" in the new physics. The American physicist and philosopher Professor Henry Margenau of Yale University, who represents the general opinion of theoreticians, insists that the waves, or rather the ψ-functions representing them, are the true reality. He maintains that this is in accord with "the philosophical view which identifies the real with the elements of experience that are causally connected in time and space." The ψ-functions are subject to causal laws in this manner through the wave equations, while the individual particles are not. Since waves determine probabilities for particles, Margenau concludes that "*probabilities are endowed with a certain measure of reality,*" for:

> Regularity is found primarily in aggregates or, when assigned to individual events, in the *probabilities* which inhere in these events. Laws govern these probabilities, they do not govern simple occurrences.

William Werkmeister, an American philosopher of science, asserts conversely in *The Basis and Structure of Knowledge* (1948) that reality is ordinarily assigned to things we can see and feel. Therefore, in the submicroscopic realm of atoms reality should be assigned to those things which interact most directly with the large-scale objects of our familiar world. It is the particles, not the waves, which produce the clicks in Geiger counters, the light flashes on the fluorescent screen and the flick of the instrument pointer. The particles are therefore real; the waves are merely a mental picture representing the mathematical equations.

Niels Bohr in developing his principle of complementarity (page 99) asserts that it is neither possible nor necessary to make a choice between waves and particles, that indeed both are essential for a complete comprehension of reality. But this point of view has again been criticized because it seems to treat particles and waves as equals when, as a matter of fact, particles are a mode of *existence* while waves are a mode of *behavior*. Margenau thus makes activity more real than existence. On the other hand, those who, following Max Born, say that photons and electrons are "really particles" must be careful not to give the word "particles" all of its classical connotations. The choice is, finally, a matter of temperament and of taste, and *de gustibus non est disputandum*.

The idea that the new physics represents a turn away from materialism is expressed in a more rational manner by the American biologist Edmund Ware Sinnott. In his book *Two Roads to Truth* (1953) he writes:

> After the revolution introduced by relativity, quantum mechanics, and nuclear physics, science was forced to modify some of its earlier conclusions. The plain truth is that the Universe is a much more complex system than it seemed to be in Newton's time. . . . Scientists accept now without surprise ideas that would have seemed preposterous not long ago. This change has been reflected in a more open-minded attitude on their part toward idealistic philosophies. For three centuries a confidently advancing science seemed to undermine the very foundations of faith, and religion was forced to modify its position in many ways or lose the support of its more thoughtful partisans. The tide, however, has begun to turn, and an aggressive idealism is going over from the defense to the attack.

That "the revolution" caused ("forced" is hardly the proper word) science to change its conclusions is true. Whether classical science "seemed to undermine the very foundations of faith" depends on what is thought to constitute these foundations. Of interest is the reference to "an aggressive idealism," for there is a special sense in which the new physics is related to idealism, one which merits some attention.

The philosophy of *idealism* is but one manner of solving the ancient metaphysical problem of the relation between an external world and our conscious internal perception of it. Idealism solves the problem by lopping off one end of it, by denying the existence of an external world. It is supported by the fact that all our awareness of the postulated real world actually occurs within ourselves; therefore, the idea that this awareness is generated by objects external to ourselves may well be an illusion.

If we look at a tree, it surely exists in our consciousness. But if we turn away, it no longer exists there; and we have no reason for assuming that it still exists anywhere. If we remember it, or if we hear some other person assuring us that it is still there, we are again experiencing nothing but mental processes within ourselves, which can afford no assurance that the tree still exists.

The common intuitive reaction to idealism is to dismiss it as absurd. The redoubtable Dr. Samuel Johnson (1709-84) thought to refute it by kicking a large rock with his toe. It has been lampooned in a doggerel concerning a certain tree in the Oxford quadrangle:

> There was a young man who said: God
> Must think it exceedingly odd
> That the sycamore tree just ceases to be
> When there's no one about in the quad.

To which God replies:

> Young man, your astonishment's odd.
> I am always about in the quad
> And that's why the tree continues to be
> As observed by
>   Yours Faithfully,
>   God.

The closing lines of this verse refer to an important concept in the philosophy of the Irish prelate George Berkeley (1685-1753), who developed the first well-elaborated formulation of the idealistic position. He accounted for the apparent permanence of many conscious perceptions, that the tree appears unchanged when seen on two successive occasions, by the assertion that it persists in the mind of God.

In spite of many attempts by competent philosophers, idealism has never been finally refuted. Since the existence of something which is not causing sensory perceptions in any conscious being is impossible to prove experimentally, it ought, in accord with scientific tradition, to be declared meaningless; and all good scientists should be idealists. In actual fact, all of classical science has been based solidly on *realism*, on the premise that an external objective universe does exist. Scientists are generally agreed, somewhat in the spirit of Descartes, that nature is not deceiving them, that their conception of a real external world is justified. It would be exceedingly disconcerting for them to admit that they are devoting all their time and effort to the contemplation of an illusion.

If the classical scientist were challenged on his belief in an objective universe, he would reply that he can always arrange matters so that his observations *do not have any appreciable effect on the observed object*. He would assert that he is actually determining what the object was like before he observed it and what it also will be like afterward.

As was made evident in the discussion of the Heisenberg uncertainty principle (page 94), this presuposition of classical science is no longer tenable. Observations do have an effect on the thing observed; and this effect cannot be reduced, even in principle, below a certain lower limit which is by no means unappreciable for the elementary constituents of the universe. This new consideration brings about a conception of the universe and the observation of it which is akin in some respects to philosophical idealism.

Classical science had assumed *a priori* that a real world exists "out there," complete and ready to be observed. The mathematical equations of classical mechanics were thought to describe what is actually *happening* in the external world of space and time. Quantum mechanics also has its mathematical equations. The fundamental difference is that they are taken to be a description of observations, of our experience of the real world. More precisely, quantum mechanical equations predict the probabilities of finding certain values of the quantities being observed. The very idea of phenomena going on apart from any observation of them is foreign to the new physics.

The difference between classical realism and the idealism of quantum mechanics has been discussed clearly by Niels Bohr. In his book *Atomic Physics and Human Knowledge* he says:

> We meet here in a new light the old truth that in our description of nature the purpose is not to disclose the real essence of phenomena but only to track down, so far as possible, the relations between the manifold aspects of our experience.

By his reference to "the old truth" Bohr is emphasizing that science has actually been concerned at all times with the description of our experience of the world. The assumption that it describes phenomena proceeding objectively, beyond our awareness of them, has in fact always been gratuitous, a circumstance which quantum mechanics has brought forcibly to our attention.

This new scientific idealism is not quite the same as that of the philosophers. It does not deny the existence of an external reality; rather, it asserts that the nature of this reality is intimately bound up with our observation of it. Metaphysically this is of great consequence; and for studies in the forefront of physics, concerned with the ultimate entities of matter, it has direct significance. But it has little pertinence in the large-scale world; for here things may well be thought to have independent objective existence much as before.

Only many years after its formulation were the deeper and more subtle implications of the uncertainty principle generally appreciated. In his 1934 presidential address, "The New World Picture of Modern Physics," before the British Association of Science, Jeans gives an interpretation widely held at that time:

> For in the old physics the perceiving mind was a spectator; in the new it is an actor. . . . The nature depicted by the wave picture in some way embraces our minds as well as inanimate matter.

This statement is compatible with the ideas expressed by Jeans in previous quotations. It implies that in the new physics there is a closer union between the observer and the objects he observes. Here this is expressed in a manner which holds close to scientific ideas; but others less inhibited by factual knowledge are moved to add their own elaborations concerning the way in which this new relation has altered our conception of the world of inanimate matter.

Bernard Bavink, a prolific German writer on philosophy of science, observes in his book *Science and God* (1933):

> Matter and its worshippers, the materialists, simply laugh at us and tell us: "Here is one single atom, the simplest one, the hydrogen atom. . . . If you can show me how I can understand this atom as the product of a merely spiritual process—I shall believe you." It seems that spiritualism can today meet this test.

The gross "worshippers" of matter are the classical physicists; and "spiritualism," by which Bavink means not the occult but things of the human spirit, is held to be a characteristic of the new physics of matter waves.

In an article published in the popular journal *Unsere Welt* (Our World) in 1933, he implies that the matter worshippers are becoming less obdurate:

> There is today within the circles of natural scientists a willingness to restore honestly the threads from these sciences to all higher values of human life, to God and Soul, freedom of will, etc.; these threads had been temporarily all but disrupted and such a willingness had not existed for a century.

Another German writer, Aloys Wenzel, goes even farther in attributing spiritual values to the world of the new physics. In his book *Metaphysics of Contemporary Physics* (1935) there appears this passage:

> This world is rather, if we are to make a statement about its essence, a world of elementary spirits; the relations among them are determined by some rules taken from the realm of spirits. These rules can be formulated mathematically. . . . We do not know the meaning of this form, but we know the form. Only this form itself or God could know what it means intrinsically.

An explanation of the reasoning which leads to effusions of this sort is given in the book *Philosophy of Science* (1957), from which some of the quotations given here are taken, by the Viennese philosopher Philipp Frank (1884-1966), a logical and lucid expositor of metaphysical problems. Commenting specifically on Bavink's remarks concerning the hydrogen atom, he says:

> From the scientific aspect, it is hard to understand why the solutions of the Schrödinger equation are more "spiritual" than the solutions of the differential equations in Newtonian mechanics. But Bavink argues by way of analogies. The solutions of Schrödinger's wave equations (ψ-functions) can be interpreted as probabilities; probabilities are however mental phenomena; hence the ψ-function is interpreted as a mental phenomenon that happens in a human mind; the hydrogen atom is described by ψ-functions; hence the hydrogen atom is a mental phenomenon and is a product of spiritual powers. The case against materialism is proved.

Such reasoning is no doubt involved in the writings of Bavink and Wenzel and others who are at pains to find in the new physics a reaffirmation of spiritual values. Inextricably mixed up with this is the further idea, hinted at in the quotation from Jeans's address, that the Heisenberg uncertainty principle establishes a more intimate connection between the external world of inanimate objects and the consciousness of the observer, a connection which causes the cold, material, objective universe to become suffused with the warm, spiritual characteristics of the human mind. To this it must be objected, in the first place, that in the large-scale world of daily experience the departure from classical behavior which the uncertainty principle permits is so infinitesimal, so far below the limits of perception, that it can offer scant consolation to the human spirit. More basic is the fact that the uncertainty principle speaks of the effects of *observations*, not of a human observer. As Bohr and others have emphasized, what the human observer can perceive directly are pointer readings, imprints on photographic plates and, in general, happenings in the world of familiar objects. He must necessarily remain remote from the events in the atomic realm, where the uncertainty principle has significance. A direct effect of the observer's "spiritual powers" on the atomic phenomena which are the real concern of the new physics is thus excluded. Indeed, in a particle accelerator laboratory the events under investigation may be detected by automatic sensing devices which convey their data to a computer where they are subjected to an elaborate analysis. The scientist who receives the computer readout is hardly closer to the particle events than is his colleague who reads the report in a scientific journal.

The Heisenberg uncertainty principle has been accorded various other metaphysical implications quite apart from its pertinence to observations. Thus, Jeans finds that "Heisenberg now makes it appear that nature abhors accuracy and precision above all things." He says:

> . . . it is almost as though the joints of the universe had somehow worked loose, as though its mechanism had developed a certain amount of "play" such as we find in a well-worn engine. Yet the analogy is misleading if it suggests that the universe is in any way worn out or imperfect. In an old or worn engine, the degree of "play" or "loose-jointedness" varies from point to point, in the natural world it is measured by the mysterious quantity known as "Planck's constant h," which proves to be absolutely uniform throughout the universe.

Jeans is correct in giving Planck's constant as the measure of "looseness"; but he does not realize that, to continue the metaphor, if the bearings were tightened (by setting h equal to zero) the engine, as explained previously (page 101), would collapse in a heap.

The uncertainty principle is given great importance in the confrontation of the old and the new physics by Erwin Canham, a contemporary American commentator on education, science, religion and other matters. His remarks are worth quoting as an example of the persistence of erroneous ideas. They appeared in 1950 in an article, "The Twilight of Materialism," published in the Boston *Christian Science Monitor*. Says Canham:

> Throughout the nineteenth and part of the twentieth century, we lived in an atmosphere of self-confident materialism. . . . It was a mechanistic world and we were sitting on top of it. We were laying hold of the apparent forces of matter, and matter was our

God. The natural scientist challenged the spiritual revelations of the Bible with doctrines of atheism and rationalism. One might almost say that this era of materialism lasted until the day the atomic bomb burst over Hiroshima. . . .

Canham thus dates rather precisely the demise of classical physics with its "self-confident materialism," which ended, however, with the coming of the new physics, for:

In the nineteenth century they [the scientists] felt that they knew all the answers, and that the universe was wrapped up in a neat mechanistic package. Today their explanations rotate around Heisenberg's self-described "Principle of Uncertainty." This is a very hopeful change.

It is not uncommon to see the uncertainty principle interpreted in this fashion, as an admission on the part of scientists that they have become uncertain of the validity of their work. While this is sheer nonsense, there is some truth, in a wider sense not specifically related to the uncertainty principle, in the statement that physicists are today less prone to make confident statements about the nature of the physical universe than they were in the previous century. Although there never was a time when "they felt that they knew all the answers," physicists in the classical period were generally convinced that the universe is comprehensible in terms of familiar experience and that a program aimed at achieving a complete understanding of it is, in principle, possible to carry out. Recent developments have perforce brought the realization that there are in the forefront of science difficult metaphysical problems concerning the nature of objective reality and subjective cognition. It is no longer clear or certain how "all the answers" are to be found.

Because of the prominence in the new physics of Planck's constant h, it has received special attention, particularly because it is the fundamental element of *action*. This term, a rather unfortunate one because of its disparate popular connotations, has a special and well-defined meaning in physics: it is the product of *energy* and *time*. (The formula $h\nu = E$ together with the relation $\nu = 1/t$, where ν is the frequency and t is the time of one cycle of oscillation, gives the result $h = E/\nu = Et$.)

Popular commentators on the new physics have been quick to draw their own conclusions about the salutary effects of action. Jan Christian Smuts, one-time Prime Minister of South Africa, in an address delivered in 1931 before the British Association for the Advancement of Science, goes easily from "action" to the more general "activity" and says:

> . . . when we make activity instead of matter the stuff or material of the universe, a new viewpoint is subtly introduced. For the associations of matter are different from those of Action, and the dethronement of matter as our fundamental physical conception of the universe must profoundly modify our general outlook and viewpoints. The New Physics has proved a solvent for some of the most ancient and hardest concepts of traditional human experience and brought a *rapprochement* and reconciliation between the material and organic or physical orders within measurable distance.

In a similar vein the British divine the Rev. J. H. Morrison in his book *Christian Faith and the Science of Today* (1936) asserts that there is a spiritual activity in all of reality and that:

> . . . all this is deeply in accord with some of the findings of the science of today. The new physics has done away with dead matter and introduced, in its stead, action as the ultimate physical reality, if physical it can be called.

Comments on these absurdities are superfluous except perhaps to point out that the importance of action is not a concept originating in the twentieth century, as these writers seem to believe. The principle of action was used by Hero of Alexandria in the first century A.D. to explain the properties of mirrors and by Pierre de Fermat in the seventeenth century to derive an important law of optics. It was applied in the eighteenth to problems in mechanics by Pierre de Maupertuis, Leonard Euler and Joseph Lagrange, and was made the basis for an important new formulation of classical dynamics in 1830 by the Irish mathematician Sir William Hamilton.

Further contentions have been advanced to show that the new physics has made the universe more amenable to esthetic, ethical and spiritual values. Thus, it is averred that atoms have been shown to be *electrical* structures and that electricity, being mysterious and nonmaterial, is therefore spiritual. The open, filigree structure of the Rutherford atom model is said to show that matter is not as solid and material as classical science had thought it to be; and this is held to be true even more emphatically for the wavelike Schrödinger model. Even Einstein's famous relation $E = mc^2$ is advanced as proof that "mass is really nothing but energy" and therefore akin to "spiritual energy."

Finally, as a refreshing change, George Bernard Shaw has one of the characters in his play *Too True to Be Good* express himself as follows concerning the metaphysical implications of the new physics, which Shaw obviously deplores:

> Newton's universe was the stronghold of rational determinism: the stars in their orbits obeyed immutable fixed laws. . . . Everything was calculable; everything happened because it must. Here was my faith. Here I found my dogma of infallibility. And now—now, what is left of it? The orbit of the electron obeys no law. . . . All is caprice, the calculable world has become incalculable. Purpose and design, the pretext of all the vilest superstitions, have risen from the dead and cast down the mighty from their seats—and put paper crowns on presumptuous fools.

With due regard for the prerogatives of poetic license, this passage sums up aptly some of the misconceptions, erroneous analogies and metaphors which have here been considered. The question arises as to how these could have become so prevalent. Part of the explanation lies no doubt in that many concepts of the new physics unfortunately can be formulated correctly only in the esoteric jargon of scientific discourse, or better, in abstract mathematical symbols. When these matters are discussed outside the purlieus of professional science, it is difficult to avoid similes and words in common usage which may easily be misleading. If later these secondhand reports are reinterpreted, at third hand, by writers who might have objectives of their own to promote, the results may well bear but dubious relations to the original circumstances.

The defense of esthetic, ethical and spiritual values against the supposedly hostile tenor of classical science is neither feasible nor necessary. Involved here are philosophical problems which have existed before the new or the old physics. There is the question of whether it is needful, possible or desirable to establish a scientific or other objective justification for the subjective feelings and emotions related to concepts of truth and beauty, harmony and justice, wonder, faith and reverence. There is the broader problem of the relations between man's rational, emotional and spiritual apperceptions and the world around him. Certainly these problems cannot be solved, or brushed aside, by recourse to one particular development in one particular area of science, a development which is still much in flux and may yet take surprising turnings.

16

Causality

CAUSALITY IN PHILOSOPHY

THE meaning of *causality*, its implications for the universe at large and for the human state, has been from remote times an important consideration to contemplative men. To the ancient Greek scholars it was a philosophical problem; in the Middle Ages it became largely a theological one. Throughout the period of classical science causality was held quite consistently by scientists to be essentially a manifestation of uniformity in nature, but it was interpreted by philosophers in widely diverse ways.

The new physics of the twentieth century has wrought changes in the scientific conception of causality, and this has brought forth renewed consideration of its implications for old metaphysical problems. To discuss these recent developments effectively, it is essential to examine carfully how causality was actually understood over the centuries both by philosophers and by scientists.

Primitive man must have been impelled to attribute causes to everything which happened around him and to him. For if these happenings were due to blind chance, he could do nothing about them; but if they were caused by friendly or hostile spirits, he could attempt to ensure himself of favorable happenings and to prevent harmful ones by performing acts designed to please or placate or drive away these spirits. No doubt he thought of causes as having an anthropomorphic character, as when he himself caused an enemy to fall by the blow of a club.

Animistic ideas persisted in the metaphysical systems of the Greek philosophers, but the peculiar Greek genius for abstract thought purged the animism of a crude anthropomorphic guise. Inanimate objects were thought to act through innate tendencies and desires, not through the mediation of manlike spirits dwelling within them. Spirits there were in profusion, and gods as well; but they were not in constant control. Indeed, the great Greek tragedies set forth the idea of a powerful fate and destiny above men and gods.

The idea of an all-pervading causal principle in the universe took various forms. For Pythagoras (582-497 B.C.), famous for his theorem about the squares on the sides of a triangle, a mystical spirit of mathematical order held sway. He saw an example of this in his discovery of a simple mathematical relation between the lengths of the stretched strings of the harp and the harmony of their tones. The idea of a musical harmony immanent in nature persisted for many centuries. Two thousand years later Johannes Kepler strove to establish in the solar system a "harmony of the planets."

Plato (427-347 B.C.), the greatest of Greek philosophers, followed Pythagoras in emphasizing the importance of mathematics; but he conceived of it as only a part of a more general system of abstract, nonmaterial ideal *Ideas*. These are models of perfection which everything in the universe, the material, the ethical and the esthetic, strives to attain. The causal principle of the universe is thus a trend toward perfect order, justice and truth, beauty and harmony in all things. This conception of an abstract ideal, of which all real objects are but approximations, did unfortunately retard progress in experimental science for centuries. It implied that true knowledge could be acquired only through philosophical contemplation of abstract ideas, not through observation of the accidental and imperfect things in the real world.

Other Greek philosophers did look about them and studied nature as best they could with their very limited resources. But they were prone to take off quickly from a few observations with sweeping metaphysical generalities. Democritus arrived at the idea of a universe consisting of indestructible atoms in a void; and Aristotle constructed it out of the "four elements," not of actual earth, water, air and fire, but of four abstract entities having properties akin to these actual realities. In the same manner he concluded that the planets, being celestial and therefore perfect, must move on perfect orbits which could only be circles, an idea which persisted tenaciously until Kepler very reluctantly had to admit in the face of inescapable experimental evidence that the orbits are ellipses.

Aristotle distinguished four kinds of causes operating in the universe: formal causes which are plans or designs, final causes or purposes, material causes which reside in matter and efficient causes which effect changes or happenings. Archimedes (287-212 B.C.), who was both a great mathematician and a scientist capable of applying his knowledge to practical affairs, emphasized a principle of causality, akin to Aristotle's efficient cause, which brings about the result that things behave everywhere and at all times in an orderly and predictable manner.

In the Hebraic-Christian world Plato's Ideas of perfect order, truth and beauty came to reside in the mind of the one perfect and infinite God. He was the lawgiver for man in the moral and spiritual realm and the source of order and harmony in the material world. The laws governing this world were held to be the direct expression of God's will and were therefore inviolable although they could be, and indeed were on occasion, set aside by God himself.

René Descartes broke away in important respects from medieval scholasticism. He considered the whole universe, except man, to be running by itself, like a machine, without the constant and immediate concern of God. Yet God's plan and purpose transcend human comprehension; therefore, mysteries can exist.

The Dutch philosopher Baruch Spinoza (1632-77), of the generation following Descartes, developed a mystical conception of causality in accord with his pantheistic philosophy. He thought of God as an all-pervading spiritual force present everywhere and at all times throughout the universe. Causality, flowing from this divine presence, could never be violated; even miracles are one of its manifestations.

Newton and his contemporaries developed the concept, which was to remain essentially unchanged for scientists over the following three centuries, that causality is inherent in the nature of the physical world itself. This idea, however, was not originated by the founders of classical science. It had been expressed well over a century earlier by that universal Renaissance genius Leonardo da Vinci (1452-1519), painter, sculptor, musician, architect, statesman, philosopher and scientist. "Nature," he wrote, "is constrained by the rational order of her law which lives infused in her."

Immanuel Kant, who was three years old when Newton died, was one of the first great philosophers to be strongly influenced by the rise of Newtonian science. At a time when the cosmological theories of Descartes were still prevalent in Europe, he advocated Newton's system of celestial mechanics and even supplemented it substantially in his scientific treatise *A Theory of the Heavens*, published in 1755. In his major philosophical work, *A Critique of Pure Reason*, he raised the principle of causality to the heights of the *a priori*, asserting that causality is a logically necessary precondition for all rational thought and therefore in no need of support by factual evidence. He defines causality thus: "Everything that happens presupposes something from which it follows according to a rule." Since a rule, a guide of regulation for conduct, may be formulated for any kind of consistent behavior, this statement is too broad to serve as a definition. In a second edition of the *Critique*, the wording is changed to: "All changes happen according to a law of connection between cause and effect," which defines causality clearly as a relation between *cause* and *effect*.

The German philosopher Arthur Schopenhauer (1788-1860) was as thorough an idealist as Berkeley. To him all objects and all happenings are *ideas*, that is, they are wholly mental. In his most important work, *The World as Will and Idea*, he elaborates the thesis that, since *willing* is the dominant mental drive, causality is a universal will in nature.

Georg Wilhelm Friedrich Hegel (1770-1831), the leading German philosopher after Kant, had his greatest interest in man and mind rather than in nature. His knowledge of science lay primarily in biology, and here he saw activity apparently directed by a *purpose* toward desirable ends. He conceived of causality therefore in a *teleological* sense: events are caused, not by circumstances in the past, but by a striving toward future goals.

Teleology, which Hegel advanced as the nature of causality in general, later became in the long controversy between the vitalists and the mechanists the primary distinction between the nature of causality in the inanimate and the animate realm. Thus, a flower is thought to acquire nectar and a bright color so that it might be pollinized by insects and produce seed. Therefore, the flower's activities are "caused" by an event which lies in the future, while the causes of events among inanimate things invariably lie in the past. Similarly, the desire to have a serene and safe old age is thought of as the "cause" of an industrious and frugal way of life in youth. Here it is clear that the *prior decision* to prepare well for later years is the cause of the *subsequent* frugality, which might well cease if the decision were changed. And, in actual fact, it is the existence of the flower which brings the seed, even as the seed brings the flower.

All of these philosophical conceptions of causality involve in various ways the idea of a mystical *nexus* whereby the cause brings the effect into being. The Scottish philosopher David Hume (1711-76) sought to purge causality of these metaphysical connotations by putting it on a thoroughly practical basis. In his important treatise on epistemology, *An Enquiry Concerning the Human Understanding*, he writes:

> The only immediate utility of all science is to teach us how to control and regulate future events by their causes. . . . Similar events are always conjoined with similar, of this we have experience: therefore we may define a cause to be *an object followed by another and where all the objects similar to the first are followed by objects similar to the second. Or, in other words, where if the first object had not been, the second never had existed.*

In this rather clumsy wording, in which "object" might be better rendered as "event," Hume is saying that a situation C and a subsequent situation E are related as *cause* and *effect*, if the occurrence of C (or a situation similar to it) is always followed by E (or something similar), and if E never occurs unless C has occurred previously. "Similar" is, of course, a vague term. Hume included it in his definition because he wanted to make causality *experimentally verifiable*; and he realized, correctly, that a given situation cannot be found to reoccur a second time if it is defined too completely and precisely. Yet his empiricism made him an avowed idealist. It was his conviction that, since all our awareness of everything external to our own consciousness comes to us via sense stimuli, there is no justification for postulating the existence of a real external world. Our sense impressions have no anchor in a permanent reality; and the fact that we have become aware of a particular causal sequence of C to E, even a very large number of times, is no proof that C will be followed by E on future occasions. He concludes that our belief in causality is no more than a *habit* which, he rightly avers, is not an adequate basis for belief.

John Stuart Mill (1806-1873), the most celebrated English philosopher of the nineteenth century, with the advantages of a hundred years of clarification of metaphysical ideas, was able to improve greatly on Hume's position. Mill is a representative of the *positivist* philosophy, which asserts that, while knowledge comes primarily through the senses, it also includes the relations which the conscious mind formulates concerning the evidence afforded by the senses. Further, positivism agrees with idealism that there is no way of proving that an external world exists, but it insists that there is also no proof that it does not.

Mill gives his conception of causality in these words:

> The law of causation, the recognition of which is the main pillar of science, is but the familiar truth that invariability of succession is found by observation to obtain between every fact in nature and some other fact which has preceded it.

Thus, like Hume, Mill makes "invariability of succession" the essence of causality, and again like Hume, he gives it an empirical ("by observation") basis. He strips causality of logical necessity and removes the idea of *compulsion*. He analyzes carefully the circumstances under which he believes that one may assume the existence of a cause-and-effect relation between two events C and E: C occurs spatially close to E, C is followed immediately by E, and C is always followed by E. He does not explicitly refute Hume's contention that causality is a habit of thought. But he does discuss the methods whereby a causal relation may be established, for example, by the *method of differences*:

> If an instance in which the phenomenon under investigation occurs and an instance in which it does not occur, have every circumstance in common save one, that occurring only in the former; the circumstance in which alone the two instances differ, is . . . the cause, or an indispensable part of the cause, of the phenomenon.

This clearly stated principle is used in many areas of science. Thus, an experiment performed on laboratory animals to test the effect of a new drug always involves *two* groups, chosen to be as nearly alike as possible in size, age, housing, feeding, and so on, with the single difference that one group receives the drug while the other, the *control* group, does not. In accord with the method of differences, any effect observed in the former and not in the latter may be fairly taken to be caused by the drug.

Many more philosophers with many yet different views could be considered. But what has been presented suffices to show that the concept of causality is, in philosophy, not a simple one and that there have been many different opinions about it. At the present time most of these opinions are being reconsidered, amended, clarified, criticized and defended; but no consensus has yet been achieved.

CAUSALITY IN CLASSICAL SCIENCE

In contrast to the situation in philosophy where, as has just been made evident, the principle of causality acquired diverse meanings, the principle's significance in science remained essentially unchanged throughout the classical period and, indeed, up to the present time. However, following the development of the new quantum mechanics, the *validity* of the principle came to be questioned in ways which have already been noted in the discussion (page 79) of electron diffraction experiments. Here and in other discussions the words "causality" and "determinism" have been employed somewhat loosely in the sense in which they are usually understood in everyday affairs, in technology and engineering and by scientists themselves. It is now in order to examine more critically how the concept of causality developed in classical physics and how it was actually understood and used. Then it will be possible to consider in some detail the ideas both of philosophers and of scientists concerning the changes wrought by the new physics.

First of all, it is appropriate to consider the relations of the three terms *causality, determinism* and *predictability*. Causality is any general principle which aims to account for the course of happenings. In some of its forms, such as the primitive idea of the sway of capricious spirits, it certainly does not connote a reliable determination of events. A deterministic causality is evident in the philosophy of Descartes, who thought of the physical universe as a great machine; but it is expressed more strongly, though in widely different ways, by Kant, for whom causality was an *a priori* necessity for all logical thought, and by Mill, who made causality "the main pillar of science." In predictability lies the practical use of determinism, as emphasized in Hume's "immediate utility of all science."

As will be seen shortly, the causality of classical science is thoroughly deterministic so that in this context the two terms are essentially synonymous. Further, while some philosophical conceptions of causality are compatible with idealism, classical science always associates causality with objective reality.

Throughout most of the classical period of science causality or determinism was closely related to the concept of the physical universe as a great machine. A machine is made of interacting parts; and the parts of the world machine were of the simplest sort: pointlike mass particles with forces acting between them. The idea of a universe built of nothing but particles goes back to Leucippus and Democritus; but Newton endowed them with forces, forces acting along the line between them and depending only on their distance of separation. This conception was raised to the position of the central concern of science by Hermann von Helmholtz, who wrote:

> Finally, therefore, we discover that the problem of physical material science is to refer natural phenomena back to unchangeable attractive and repulsive forces between bodies whose intensity depends wholly upon distance. The solubility of this problem is the condition for the comprehensibility of nature. . . . And its [science's] vocation will be ended as soon as the reduction of natural phenomena to simple forces is complete and the proof given that this is the only reduction of which phenomena are capable.

It is fortunate for his scientific reputation that Helmholtz left an implied possibility open that this might not be "the only reduction." For even as these words were written, evidence was appearing that it is not possible to account for all phenomena in terms of particles interacting by simple forces. The decline of the wholly mechanistic physics had indeed begun some years earlier with Michael Faraday's postulated lines of electric and magnetic force made of he knew not what but certainly not of particles. With Maxwell the lines of force became electromagnetic fields, first tied to electrically charged particles and later traveling in wavelike fashion out into space, devoid of any material substance.

The inroads made on the mechanistic conception of nature by these new developments are evident in the plaintive remarks of Lord Kelvin (1824-1907), a leading figure in English scientific circles during the latter half of the nineteenth century:

> I am never content until I have constructed a mechanical model of the object I am studying. If I succeed in making one I understand; otherwise I do not. I wish to understand light as fully as possible, without introducing things that I understand still less.

In spite of these vicissitudes, the principle of causality was never questioned. Toward the end of the nineteenth century it was raised among scientists to the eminence of a self-evident truth, even as Kant had done a century before on metaphysical grounds. Typical of this attitude is this statement of Boltzmann, given in his *Physiologische Optik*:

> The causal law bears the character of a purely logical law even in that the consequences derived from it do not really concern experience itself but the understanding thereof, and that therefore it could never be refuted by any possible experience.

In view of what has happened in the meantime, this statement exemplifies the hazards of making predictions about the future course of science. Maxwell was more circumspect:

> The promotion of natural knowledge may tend to remove that prejudice in favour of determinism which seems to arise from assuming that the physical science of the future is a mere image of the past.

The "image of the past" did hold well beyond Maxwell's time. Although his work made a naïve mechanistic causality untenable, the broader concept of a causally determined universe could still be maintained. It is instructive to compare this scientific concept with that of the philosophers.

Philosophical discussions of causality are, in general, much concerned with events. This concern is shared by experimental scientists, whose work consists principally of observing the sequence of events, by engineers, who apply scientific knowledge to the control of events, and by everyone in the conduct of many everyday affairs. Those, however, who work in the forefront of science, seeking to formulate general principles and laws, rarely deal with actual events in nature. One reason is that most events are too complex.

Consider, for example, an apparently simple occurrence, the flight of a projectile. The paths of projectiles were studied carefully by natural philosophers of the Middle Ages, for even before the invention of gunpowder, when military fortifications were demolished by huge boulders tossed by catapults, it was thought advantageous to hit the intended target. But detailed analysis showed that the motion of a projectile is exceedingly complex; at every instant during its flight it changes both the magnitude and the direction of its velocity. The problem remained intractable until Galileo, in the seventeenth century, had the brilliant idea of taking the motion apart into its *component motions,* a vertical (up and down) and a horizontal (straight ahead) motion. These separate components proved readily comprehensible, and thus the matter was solved. This is typical of the way in which science proceeds; it deals, not with actual events, but with the simpler components into which it succeeds in separating them. Thus, the action of a muscle had to be resolved into its mechanical, anatomical, physiological and biochemical aspects before an understanding could be achieved.

There is a far more significant consideration regarding the scientist's concern with events, one which determines largely his conception of the causality principle. This is indicated in the following passage from a paper, "On the Notion of Cause," by Bertrand Russell, English mathematician and philosopher and recipient of the 1950 Nobel Prize in literature:

> All philosophers, of every school, imagine that causation is one
> of the fundamental axioms of science, yet, oddly enough, in
> advanced science, such as gravitational astronomy, the word
> "cause" never occurs. . . . The Law of Causality, I believe, like

much that passes among philosophers, is a relic of a bygone age, surviving, like the monarchy, only because it is erroneously supposed to do no harm.

This faintly satirical passage is certainly correct in its statement that "in advanced science . . . the word 'cause' never occurs"; but it goes too far in its implication that causality is for science "a relic of a bygone age."

The meaning of Russell's words may be made clear by considering his example of "gravitational astronomy." When Newton began his study of gravity, he was surely interested in *events*, in the fall of an apple, in the motion of the moon about the earth, and in their *causes*, much in the sense of Hume and Mill. But after he had brought his work to the state of an "advanced science," events retreated into the background, and there remained, as the central item of interest, the universal law of gravity. This expresses in a mathematical formula the general *relation* of the *distance* between two objects and the *force* of gravity acting upon them. Here there is no longer any specific reference to causes or to temporal succession of *before* and *after*, which is so prominent in philosophical conceptions of causality. Nor is there any criterion for singling out either force or distance as the cause and the other as the effect.

This idea of *relation*, central to the scientific meaning of causality, is well worth examining by a second example. Consider a small, air-filled, sealed rubber balloon held between the cupped hands. If it is squeezed to a smaller volume, the air pressure within rises. For this particular event it seems reasonable to think of the squeezing as the cause and the resulting pressure rise as the effect. But a scientist desiring to study the behavior of gases might set up an apparatus for compressing the balloon mechanically, equipped with two instruments, one indicating the volume of the balloon in cubic inches, the other the air pressure in pounds per square inch. He will then find (if he is careful to maintain the temperatures of the air in the balloon constant for all his readings) that, as he decreases the volume successively from 100 to 50 to 25 cubic inches, the pressure rises correspondingly, from 20 to 40 to 80 pounds per square inch. That is, every time the volume is halved, the pressure is doubled; or more generally, the pressure rises in inverse proportion to the decrease in volume. Briefly, in the language of mathematics, he has found the relation:

$$p = C/v$$

where C is a constant depending on the conditions of the experiment. (This is the mathematical expression of Boyle's law, page 69.) In the case just considered, $C = 2000$, since the balloon was filled with air so as to make its pressure 20 at the initial volume of 100.

The particular events of this or of any number of similar experiments are of no further interest. Again with regard to the general relation here discovered, it is meaningless to distinguish between cause and effect. The experimenter might manipulate the apparatus so as to change the pressure over a predetermined sequence of values and read the resulting values of volume, or he might proceed the other way about. The relation of p and v is the same in either case.

The scientific conception of causality is expressed best by an expression taken from mathematics: the *functional relation between variables*. Variables, or variable quantities, are entities, such as pressure and volume in the balloon experiments, which vary or are varied in a given situation. To say that two variables stand in a known functional relation to each other means *only* that, if one is given, the other may be found. These relations are frequently expressed as mathematical formulas, such as the one relating pressure and volume, which for a given situation (a given value of C) affords a value of p corresponding to any value of v.

Functional relations may be expressed in various other ways, as by a graph or a table. With a table listing the numbers of inhabitants in a hundred large cities, the name of the city may be found if the number of inhabitants is given, and vice versa. Obviously, a functional relation carries no implication of cause and effect in the metaphysical sense.

A good part of the business of an exact science is to establish functional relations between variables. But it is not true, as Russell implies, that causal relations for events have been banished from science. Such relations are scrutinized precisely and in detail in all of experimental science, and they are considered carefully in the building of new theories. But they are only a means toward an end. Those who work in "pure" science and in the forefront of the advance are interested primarily in discovering functional relationships.

When a relation of this kind has been found to be broadly valid and to express an important fact about the workings of the physical universe, it acquires the stature of a *law* of nature. The principle of causality in science may be said to reside simply in the constancy and reliability of its laws. With due regard for the two facts that the experimental data upon which laws are based are never ideally precise, and that all theoretical relations are tentative and subject to revision by new discoveries, scientific causality connotes no more and no less than the *uniformity of nature*.

An "event" in the scientific sense is an observable state of affairs together with the causal relations which are pertinent to it. The conditions obtaining at the start of an experiment or a set of observations are known as the initial conditions. In an observation of the solar system, say, for the purpose of predicting an eclipse of the sun, the *initial conditions* would be given as positions and motions of the sun, moon and earth. These, together with the Newtonian laws of motion and of gravitation, constitute the closest analogy which science affords to the philosophical idea of a "cause"; and the eclipse is the "effect." It is, however, true that this same cause could be used to determine the occurrence of an eclipse at a former date in history, an event which neither Hume nor Mill would consider an effect.

Scientific causality, in relation to events as just described, results in *determinism*, it being understood that to say that event B determines event A means simply that, given B, it is possible to *calculate* A, and vice versa. Thus, with regard to the way in which causality is *used* in an exact science, it may be characterized by the statement: given the condition of a system at one particular instant, it is possible to determine by calculation its condition at any other, past or future, instant.

It is noteworthy that the scientific conception of causality is completely devoid of any connotation of coercion or *compulsion*. This is well worth stressing because it seems to be easily forgotten or misunderstood in discussions of such matters as free will and moral responsibility. This is understandable, for mankind has been

conscious of its condition for a hundred millennia and philosophers have been discussing it rationally and systematically for
twenty centuries. But science in the modern sense is barely three
hundred years old. It is hardly surprising that a conception which
science has formulated so recently for its own use has not yet
become a part of our intuitive heritage.

The scientific meaning of causality seems clear and consistent.
Yet, as the Viennese philosopher Moritz Schlick, among others,
has pointed out, it is not free of difficulties. Schlick thinks of the
scientist observing a part of the universe as it pursues its course
through space and time. To test whether the principle of causality
holds, the scientist makes a series of observations and then tries
to find a law, a mathematical formula, which is in accord with
these observations. If he succeeds, he concludes that the universe
is a causal one. This procedure, says Schlick, is meaningless, for a
sufficiently astute mathematician can devise a mathematical
formula to fit any sequence of observations whatsoever. Therefore,
the scientific test is no test at all.

Scientists have been well aware of this difficulty, and they have
met it by placing restrictions on the kind of mathematical
formulas which may be accepted as laws. The requirement is
essentially that they be "simple"; and scientists are generally
agreed that they know what this means and that, as a matter of
fact, the basic laws of science *are* simple. But Schlick insists that
simplicity is a vague concept and not admissible as a criterion
for causality. He suggests that a causal law be required to fit not
only the observations, already made, from which it is deduced,
but others, made later, which were not used in its derivation.

Schlick is right in emphasizing the importance of *extrapolation*,
a kind of prediction out into the unknown; for the power of laws
to suggest new areas of experimental investigation has contributed
greatly to the progress of science. But he does not grasp correctly
the way in which scientists actually put causality to the test.
It is true, as Schlick sees it, that the universe moves as one
indivisible whole along a path of no return and that, in principle,
any sequence of events happens only once. Every part of the
universe is constantly under the influence of every other part.
The fall of an apple is affected by the gravitation force, not only
of the earth, but also of the moon, the sun and all the planets,

and indeed of every star in the heavens; and all of these will never again return together to their former positions relative to the apple. But the scientist is not quite willing to adopt, as a working principle, the poetic hyperbole:

> Thou canst not stir a flower
> Without disturbing a star.

He is certain that he can *isolate* a portion of the universe sufficiently so that the disturbing effects of all else are well below the limits of his experimental precision. He can then bring this limited system to the same initial state over and over again and observe whether the same sequence of events occurs in each case. If so, he derives a mathematical relation to fit this sequence, a relation which he can later use to *calculate* the events following from this initial state. Causality is thus demonstrated, and the advantages of predictability are achieved.

Yet even this is not the way in which causality is really made evident. As already indicated, scientists are not interested primarily in producing like sequences of *events* under like conditions but in finding uniformity of *relations*. Considering again the experiment of the air-filled balloon, the simple pressure-volume relation found there has actually been established as a broadly valid law, not by repeating a single experiment under the same conditions, but by performing many different experiments, using samples of various gases under widely varying conditions. To obtain consistent results in such a group of experiments, it is still necessary to isolate the observed systems carefully from the disturbing effects of the rest of the universe.

Science has evolved a conception of causality which is based upon its own well-defined procedures and is well suited to its needs. Nevertheless, metaphysical questions have a way of intruding. To thoughtful scientists it has been a constant source of wonder that nature shows such a large measure of correlation with their simple mathematical formulas. "The eternal mystery of the world," said Einstein, "is its comprehensibility."

The question has been debated endlessly by philosophers and scientists alike as to whether the uniformities expressed by scientific law reside in nature and are *discovered*, much as an explorer discovers new lands, or whether they are *invented* and applied to nature by the mind of the scientist. Kant, who placed the formulation of the principle of causality in the mind, ascribed all the apparent order in nature to this same source. "The understanding," he wrote, "does not draw its laws from nature but prescribes them to nature." Sommerfeld, in common with many other scientists, felt that there is an intolerable arrogance in the thought of "prescribing" to nature. Rather, scientists in a spirit of humility should hope that through unremitting labors they might achieve some small measure of comprehension of her wonders.

The uniformities of nature do not lie open for inspection. They must be framed somehow out of the welter of observed complexities. And this involves value judgments which cannot be made according to set rules of scientific procedure. Ever since the scholastic philosopher William of Ockham (1270-1347) proposed the principle of "Ockham's razor," the shearing away of nonessentials, *simplicity* has been urged as a criterion in the formulation of concepts and laws. Scientists have preferred certain forms of laws for other nonscientific reasons: because they are "elegant" or "beautiful."

In this century the universe has been extended outwardly from our own galaxy, the star group of our Milky Way, to enormously greater reaches of space and inwardly from the periphery of the atom to entities a million times smaller. The wonder is that the human mind can conceive at all of these realms so remote from the environment in which it had its slow development over a period of a thousand centuries. In his book of essays *Out of My Later Years*, Einstein, after expressing his opinion that scientists do poor philosophizing and that it might seem proper to leave this activity to philosophers, continues:

> At a time like the present, when experience forces us to seek a newer and more solid foundation, the physicist cannot simply surrender to the philosopher the critical contemplation of the theoretical foundations; for, he himself knows best, and feels more surely where the shoe pinches. In looking for a new foundation, he must try to make clear in his own mind just how far the concepts which he uses are justified and are necessitated.

17

Free Will

THE problem concerning the freedom of the human will arises from the incompatibility of the very ancient idea that there is in the universe a principle of causality to which all happenings are subject and the equally ancient intuitive conviction that our choices and decisions are made freely. For it would seem that, if the volitions leading to our actions are *caused*, they cannot be freely and spontaneously our own. It is then unreasonable that we should be held responsible for our acts or that we should be expected to have feelings of guilt or remorse for our wrongdoings. Thus, the free will problem is closely linked both to ideas of causality and of ethics and morals.

Socrates (470-399 B.C.) has been called the earliest of the great ethical philosophers although he never built a formal system of ethical doctrines. To him philosophy was not a subject to be discussed in abstract terms but to be used in the affairs of daily living both as an individual and as a citizen of the state. His own life was an exemplar of the highest ethical conduct. At the end, when he had been condemned to death by an Athenian court of law on charges brought by influential men who felt themselves personally affronted by his teachings, he refused to take advantage of an opportunity for flight provided by his loyal friends, because he believed that it is the duty of a citizen to abide by the laws of the state.

Plato, who was the first philosopher to develop a well-conceived system of ethics, was strongly influenced by the teachings of Socrates and by the example of that great philosopher's life. More specifically, Plato's ethics was related to his metaphysics, to his doctrine of ideal Ideas exerting their influence both on material things and on the minds of men. Wisdom and goodness lie in acting in accord with the ideal principles of ethics inherent in the universe itself. In Plato's philosophy the question of freedom of the will is never faced explicitly, for the good man, living in harmony with the universe, is free in the sense that he never encounters moral conflicts or the need for moral choices and never experiences a sense of compulsion of his volitions.

Aristotle, Plato's most illustrious pupil, follows him in his con ception of an ethical universe. But he emphasizes the importance in man of *reason*, which directs him to the good and the morally excellent and thus brings him happiness and tranquillity. Virtue can become a habit so that good deeds are done almost instinctively. But in general, man is conscious of his acts and is morally responsible for them. The tragic hero, though better than the run of men, is still in part to blame for his misfortune, for it is invariably a flaw in his character which strikes him down. Aristotle teaches that man's will is free in the sense that his acts are not under compulsion by forces outside himself but are caused by his own nature.

Epicurus (342-270 B.C.), who founded a famous school of philosophy at Athens, followed Democritus in his atomistic-mechanistic view of the physical universe. In his metaphysics he opposes the Platonists; he asserts that the road to truth lies in observing the real world, since all knowledge comes to us through the evidence of our sense perceptions. He does not deny the existence of the gods but believes that they lead a happy life quite free of any concern about the affairs of men. Ideas and emotions, in particular ethical principles and concepts of good and evil and of vice and virtue, exist only in the human mind and have no counterpart in the universe at large.

The Epicureans are best known for the central doctrine of their philosophy that *pleasure* is the primary object of life. Pleasure is to be understood as a tranquil state of mind, acquired by shunning both indulgence in sensual delights and asceticism and by practicing moderation in all things. Virtue is esteemed, not as an end in itself, but as a way toward happiness.

Contemporaneously there arose in Athens the Stoic school of philosophy founded by Zeno of Citium (died 264 B.C.), not to be confused with the earlier Zeno of Elea, famous for his paradoxes. The Stoics believed in the existence of a universal law or universal reason pervading the whole universe. All things act naturally in accord with this *anima mundi*, but only man can do so consciously and voluntarily. If he lives according to the fundamental principle of his nature, in harmony with the universal reason, he escapes the sway of emotions and the temptations of pleasure. Reason leads to wisdom and wisdom chooses virtue. The wise and virtuous man, knowing that all happenings are in harmony with a universal law, accepts them without complaint. Virtue is the only good, vice is the only evil. Health, riches and power as well as pain and poverty, or even death, are all matters of indifference.

Both the Stoics and the Epicureans strive to achieve a harmonious way of life, unperturbed by chance events and free of ethical discord. To this end the Stoics counsel the suppression of all worldly desires, the Epicureans their wise regulation. These two movements represent the first serious break in what had been up to the time of Aristotle an essentially unitary development of Greek philosophic thinking. In the ethical realm their dichotomy lies in that the Stoics see a harmony with human ethical concepts in the universe as a whole while the Epicureans make ethics a uniquely human concern.

Marcus Tullius Cicero (106-43 B.C.), the Roman orator, poet, philosopher and statesman, was the first to bring free will to the fore as a serious philosophical problem. He insisted that an all-pervading causal principle in the universe, such as Plato's supranatural Ideas or the universal reason of the Stoics, cannot be reconciled with individual free will and moral responsibility. Since he was convinced that these are essential both for a virtuous personal life and for the survival of the state, he was forced to deny the existence of a universal causality extending its sway over man's moral life. The concern of those who shared his views was not primarily that an external determination of moral actions would remove all reason for praise or blame, but that man is an isolated moral creature in an amoral world and must build him-

self a structure of ethical principles without the support of anything immanent in nature.

Cicero's philosophical writings had a profound influence upon
the thinking of the early Christian theologians. With them it
became a problem concerning the nature of God. If God is
omniscient and has foreknowledge of all happenings, man is a
mere puppet in his hands. Since God cannot foreknow what does
not happen, all man's volitions and actions are foreordained. He is
not responsible for his sins, and God is not justified in meting out
punishment.

St. Augustine (354-430), the most celebrated of the early church
fathers, struggled mightily with this problem. In his *Confessions,*
one of his earlier works, he writes:

> And I strained to perceive what I now heard, that free-will was
> the cause of our doing ill, and Thy just judgement, of our suffer-
> ing ill. But I was not able clearly to discern it. . . . I knew as well
> that I had a will as that I lived: when then I did will or nill any-
> thing, I was most sure, that no other than myself did will and
> nill. . . . But again I said, Who made me? Did not my God, who
> is not only good but goodness itself? Whence then came I to will
> evil and nill good, so that I am thus justly punished? Who set
> this in me, and ingrafted into me this plant of bitterness, seeing
> that I was wholly formed by my most sweet God?

To St. Augustine the free will problem is a part of the greater
one of the existence of evil in a world created by a good and just
God. Through this problem he was for a time impelled toward a
form of predestination. Because of the sin of Adam all men are
sinful and deserve eternal damnation. But some are saved by
divine grace, not merited but freely bestowed. In this he sees no
injustice. But by the doctrine of man's sinful nature he does deny
real freedom of the will.

Later in his most important work, *The City of God,* on which
he labored for fifteen years, he achieved what was to him a true
solution of the free will problem. This is set forth in these
passages:

> But it does not follow that, though there is for God a certain
> order of causes, there must therefore be nothing depending on

the free exercise of our wills, for our wills themselves are included in that order of ·causes which is certain to God, and is
embraced in this foreknowledge, for human wills are also causes
of human actions; and He who foreknew all the causes of things
would certainly among those causes not have been ignorant of
our wills.

For man does not therefore sin because God foreknew that
he would sin. Nay, it cannot be doubted but that it is the man
himself who sins when he does sin, because He, whose foreknowledge is infallible foreknew not that fate, or fortune, or something else would sin, but that the man himself would sin, who,
if he wills not, sins not. But if he shall not will to sin, even this
did God foreknow.

Thus, St. Augustine concludes, God's foreknowledge does not
conflict with man's free will. For free will does not mean that acts
are uncaused but that they are caused by the self. God knows
this self more completely than the self does. He has foreknowledge
of how the self will act out of its own nature or free will, not of
how it will be forced to act.

Although this conclusion satisfied St. Augustine, it did not end
theological discussion of free will. St. Thomas Aquinas (1225-74),
acclaimed as the greatest philosopher of the Roman Catholic
church, was a leader of the thirteenth-century scholasticism. He
taught that the will is controlled by the intellect, for the will
naturally chooses what the intellect deems most worthy. He
follows St. Augustine in the doctrine that man is by nature sinful
through the fall of Adam. But man can achieve goodness through
divine grace which becomes the sufficient cause of his actions.
He does, however, possess free will, for if he does not choose to
act in harmony with divine grace, it is not efficacious.

Jonathan Edwards (1703-58), the celebrated divine of colonial
days, who left his parish in Northampton, Massachusetts, and
removed to the frontier settlement of Stockbridge because his
theological views clashed with those of his parishioners, is generally considered the first important American philosopher. In his
book *The Freedom of the Will*, he achieves a significant advance
toward clarifying some of the conceptions of the free will problem,
being far ahead of his time in giving thought to matters of
semantics. He defines "freedom" as "the power, opportunity, or

FREE WILL 269

advantage that anyone has to do as he pleases"; and he adds that "in the propriety of speech, neither liberty nor its contrary can be properly ascribed to a thing . . . as is called the will." He insists that it is properly the *man*, not the *will*, which is free.

Through lengthy arguments Edwards arrives at the conclusion that the question whether the will is free *or* caused is not a proper one, for it is *both*. Man makes freely, without any sense of coercion, those choices which are determined by his own desires, by what seems to him to be to his best advantage. Yet everyone's judgments and decisions, for good or for evil, are determined finally by his nature, which is God-given. Indeed, says Edwards, it would be an affront to God to maintain that man might make his own decisions, quite independent of His omnipotent will.

The ideas concerning the freedom of the will held by various leading philosophers throughout the seventeenth, eighteenth and nineteenth centuries are naturally related to their conceptions of causality. Descartes, whose primary interests lay in mathematics, science and metaphysics rather than in theology, maintains that the course of all events is foreordained by God; and he also concedes that man experiences a strong sense of free will and moral responsibility. But it is neither possible nor necessary for man to attempt a resolution of the resulting problem of incompatibility.

To Spinoza, for whom a pantheistic God is a pervasive causal principle acting in and through all things, an individual personal free will apart from the will of God is unthinkable. He teaches that, insofar as we concern ourselves with material and transitory matters, we are slaves to desires and our will is frustrated. But when we turn our attention toward increasing our comprehension of God's will, our own willing becomes more harmonious with His. A complete understanding of God's will is impossible to achieve, but the striving for this understanding is our only source of such freedom as we may attain.

Hume believes that our volitions are always controlled by our emotions, for it is the emotional part of our nature which envisions the ends we wish to attain and the value to be attached to them. But reason serves to make emotions effective by directing them toward ends which are real and attainable and by selecting effective means of attaining them. Yet our emotions spring from the permanent base of our character so that our acts are finally caused, in Hume's simple and direct sense, by our natures; and we are responsible for them.

To Kant the will is related to the categorical imperative (the command which admits of no evasion) of living in accord with the moral law of duty. This is for everyone a law referring to his own nature; there is no moral value in doing one's duty through coercion, even through divine commandment. The will is free only if it is in accord with this individual law of duty.

Freedom can exist for Hegel even under apparent compulsion. In the very common circumstance of having to work at a routine job every day, a man may feel that he is trapped in a chain of events not of his own doing. But if he can identify himself as part of a whole economy, which is acting freely to achieve its ends, he comes to think of his relation to society as a mutual one wherein he is determining as much as he is determined, and he feels free. Similarly, when we observe causality operating in various parts of the material universe, it seems that all is determinism and necessity. But looking at the whole, we see that it is self-determining and therefore free. Thus, man, as he thinks of himself as a part of the whole, can feel that he is acting freely in a causal world.

Mill comes to this same conclusion, but by reasoning close to that of Hume. Being a thorough believer in universal causality, he cannot admit that volitions are free in the sense of being uncaused. Rather, the will has its causes in character and drives. These, says Mill, are largely the result of education and environment, an idea in harmony with his belief in the value of the social sciences.

The ideas of the prominent American philosopher William James, brother of the novelist Henry James, are part and parcel of his strongly pragmatic philosophy; his main concern is that a man be able to act effectively in society. He considers the world neither wholly good nor wholly bad but certainly in need of improvement. The will, he is convinced, is free, and this places a moral obligation on everyone to attempt to improve the lot of his fellow men. He believes that, while much of our character and our interests and drives are determined by inheritance and environment, there remains a certain spontaneous freedom of choice through which we control our acts and become masters of our fate.

John Dewey (1859-1952), who ranks with William James as one of the two leading American philosophers of the early twentieth century, was born and bred in New England and started his professional career as a high school teacher, which accounts for his interest in community life and in education. His ideas concerning free will are similar to those of Hegel: we are free if we feel ourselves to be part of the course of events around us and in some measure in control of them. To this he adds a pragmatism akin to that of James, for he sees this power of control as an opportunity and as a moral obligation to work toward socially desirable ends.

The ethical implications of free will are stressed in another manner by the twentieth-century American theologian Thomas Merton. In his book *Seeds of Contemplation* he writes:

> Freedom does not consist in an equal balance between good and evil choices but in the perfect love and acceptance of what is really good and the perfect hatred and rejection of what is evil, so that everything you do is good and makes you happy, and you refuse and deny and ignore every possibility that might lead to unhappiness and self-deception and grief. . . . Only the man who has rejected all evil so completely that he is unable to desire it at all is truly free.

This is certainly a forceful statement that volitions are based on character.

More names could be added and more opinions presented. But from those given, it is evident that philosophical ideas about the free will problem are numerous and diverse. With a few notable exceptions there is, however, a general consensus, expressed more or less explicitly and in a variety of ways, that our wills are free in that they spring from our own natures and are not under compulsion from anything alien to ourselves.

This was asserted a long time ago by Heraclitus of Ephesus (540-475 B.C.), who lived at the dawn of the glory that was Greece. Very little about his life is known, and only fragments of his work survive. He probably was more akin to a wandering commentator, making pithy remarks about the human condition, than to the conventional concept of a philosopher sitting with chin in hand and corrugated brow. He emphasized the existence of perpetual change in the universe and held that changeable fire is the primary substance. Yet he taught that the intellect can discern a changeless pattern in the course of events. "Over those who step into the same river," he said, "different and ever different waters flow." The waters change, the river remains unchanged. In man there is, similarly, a steadfast character and purpose which determines the course of his life: "A man's character is his destiny."

THE NATURE OF THE WILL

HAVING discussed at some length pronouncements of philosophers, theologians and scientists, it is now in order to examine the will itself and the various considerations which bears upon its nature and its freedom.

There is first of all the question of whether volition is to be thought of as a purely *mental* phenomenon or whether the overt activity resulting from volition is to be considered an inseparable part of it. Philosophical discussions do not always make a clear distinction in this matter. To the layman, just sitting and thinking seems of little avail; it is the *doing* which is important. Furthermore, behavior is the only clue to the volitions of others. Each one of us can have direct knowledge only of his *own* thoughts.

In the essay entitled "Freedom of the Will" in the book *Readings in Philosophical Analysis* (1949), by Herbert Feigl and Wilfred Sellars, the layman's conception of free will is presented thus:

> The layman, or "the man in the street," may well wonder why anyone should ever have denied the existence of free will, since the difference between freedom and compulsion appears to him to be a fact that he finds exemplified every day. . . .
>
> I am free (he would say) when I have the power to abstain from an intended action, when my actions are under my control. I act under compulsion when I am forced to do a thing I had no intention of doing. Freedom, therefore, implies the existence of alternatives, any one of which I could have chosen had I so

desired; compulsion implies the removal of one or more of these alternatives. . . . My conduct is free (so the layman might conclude) only so long as my actions are not determined by a "must."

L. Susan Stebbing, Professor of Philosophy at the University of London, gives some reasons for considering actions as related to the free will problem. In her book *Philosophy and the Physicists* she writes:

> For the world in which we act is one in which other persons also act, and their actions conflict with mine, thwart my desires, and bring me unmerited suffering. . . . Does not the sting of the problem lie there? It stings because sometimes I feel helpless, feel compelled, feel useless, feel that however much I try and whatever I may do, I cannot do what I want to do although I know that what I want is worth while.

These statements clearly imply that the freedom of the will is bound up with matters which affect *actions*. Of course, consciously willed and desired actions are hedged about by a variety of constraints, physical, economic, legal, conventional and ethical. There are desires, say, to soar up into the sky unaided, which are physically impossible to satisfy; there are others, as the ambition to excel in some sport, which require great effort; still others which, like the urge to rob a bank, involve great risk of punishment; and finally there are those, such as wishing to make a modest contribution to a charity, which may be carried out freely. It would seem that the consideration of such matters complicates the problem endlessly without getting to the nub of it, for volition is essentially a process going on in the *mind*.

In recent times the free will problem has been brought into the realm of psychology, where from the scientific point of view it indeed belongs. Psychologists have done much to clarify the concept of volition by considering it apart from actions, except insofar as overt behavior affords insights into what is going on in the mind. The process of volition has been analyzed into desires, motives and drives, both conscious and subconscious, mental effort to reach decisions, and the final making of a choice. Further studies have dealt with effects of external influences, and with ethical questions of responsibility, guilt and remorse.

Psychologists generally consider volitions to be free when they appear related to one's own character, desires, motives and intentions, and as not free when they are the result of external influences and manipulations. The latter is particularly true if a certain volition, as evidenced by an observable act, can be elicited in a uniform and predictable manner by a given external influence. It would seem, in general, that any happening which can be made to occur uniformly and predictably by a given manipulation, as the ring of a doorbell by the push on the button, cannot be attributed to free volition.

Upon closer scrutiny this apparently clear distinction between free and controlled volitions is seen to be fraught with difficulties. Thus, if a certain volition occurs as the result of advice or persuasion, there is a question of whether it should be attributed to alien manipulation or to a change of character produced by the advice. Involved here is the broad and oft-debated question of "nature versus nurture," of the relative effects, upon ourselves and all our activities, of our innate, inherited characteristics and of our environment.

Observations and discoveries in the life sciences of neuroanatomy and neurophysiology have important implications for the nature of the will. Here a broad distinction can be made between the *involuntary* or *autonomic* nervous system and the *voluntary* system, which is concerned with conscious behavior. Many of the activities going on in our bodies under nerve control proceed without our awareness. These are the functions which, in the words of the French physiologist Claude Bernard (1813-78), "nature thought it prudent to remove from the caprice of an ignorant will." They include visceral functions, such as the flow of digestive juices when food enters the stomach, and the maintenance of normal body temperature and blood composition.

Many other bodily functions are involuntary, as the contraction of the pupil of the eye in bright light and the leg jerk occurring when the physician "tests the reflexes" by a tap with a small rubber hammer just below the kneecap. The former is an example of an automatic response which happens without our knowledge; the latter is one of which we are aware although we have no control over it. There are still others, such as sneezing and coughing, which are difficult but not impossible to control. But all of these, to the extent that they are automatic and predictable responses to given stimuli, are certainly not manifestations of a free will.

Many muscular actions, such as walking, maintaining equilibrium and breathing, are ordinarily involuntary but can be brought instantly under conscious control. Others, such as writing, painting or singing, are nearly always voluntary. Yet there is no sharp distinction. Indeed, many activities pass gradually with repetition from a consciously controlled to a nearly automatic stage. The first attempts at playing the piano are made consciously, finger by finger. But the accomplished pianist is conscious only of the general effect he wishes to produce; the fingers carry out the details of motion "by themselves." Finally, there are the wholly mental functions, sensory perceptions, memory, reasoning, emotions, value judgments.

Broadly speaking, there are relations between these various kinds of physiological nerve activity and overt behavior and the anatomical structure of the nervous system and brain. Those activities which are most complex, which appeared latest in the process of evolution and are most highly developed in man, are concentrated in the uppermost part of the brain, the *cerebral cortex*. More primitive functions, particularly the autonomic and involuntary, are related to parts located lower and nearer to, or in, the spinal column.

The divisions are by no means simple and clear; the interconnections of nerve pathways are numerous and enormously complex. When a finger touches a hot stove, there is an immediate quick nerve transmission along a path from the sensory nerve in the finger to the arm muscles, which jerk the finger away; but a fraction of a second later impulses have arrived at the cortex, which engender sensations of surprise, pain and annoyance, and perhaps an outcry. Also, conscious thought and emotions affect breathing, heart rate, the viscera and the blood vessels, as in blushing; and the unperceived states of bodily functions exert an influence upon higher mental activities.

Neither in the anatomical structure of the nervous system nor in its physiological functioning nor in overt behavior is it possible to draw a sharp line of demarcation between involuntary and conscious, voluntary activities.

Indeed, some neurophysiologists and psychologists assert that there is no difference in kind but only in complexity between such events, for example, as the response of an enthusiastic audience to an outstanding artistic performance and the leg jerk due to the physician's hammer tap. *All* responses have this in common: they are conditioned *jointly* by the nature of the stimulus and the state of the living organism upon which the stimulus acts. It is asserted that the jerk of the leg and the applause are, in principle, equally automatic and predictable, though both the stimulus and the system which receives it are in the latter case vastly more subtle and complex and impossible to comprehend completely. Nevertheless, if one knows that a friend is an enthusiastic admirer of Mozart's music and knows also that his medical history is normal, one would probably be willing to wager a larger sum that he will applaud a superb performance of *The Magic Flute* than that his leg will jerk in response to the knee tap.

From such considerations it would follow that the "nature" or "character" to which so many philosophers lay volitions involves physiological as well as psychological characteristics. Further, it appears that the question of freedom of the will, even of the extent to which it lies in the realm of the conscious, is not answered easily.

There are still other observations which imply that our estimation of the freedom of the will is exaggerated. A very prevalent opinion concerns the possibility of control and the predictability of the volitions of *others*. Each of us is convinced that his own decisions, regarding, for example, the purchase of an automobile, are made freely, after intelligent consideration of various alternatives. But we are not so sure of our neighbor. We suspect that some of his choices may have been affected by external manipulation, a suspicion which the advertising industry endorses to the tune of many millions of dollars annually. It is well known that an advertisement which supplies factual information to aid in an intelligent free decision is generally less effective than one which appeals directly to the emotions and is designed to manipulate the decision in a predetermined desired direction.

Predicting the behavior and thus the volitions of others is very common and to a large extent successful, both individually and *en masse*. If the prediction fails, we are prone to lay this to insufficient information rather than to unpredictability in general.

Insofar as the success of prediction is based on knowledge of individual traits or of "human nature," it reaffirms the idea that volitions stem from character. But predictions based upon known effects of external manipulations are a clear disproof of free will. Unfortunately, such control of volitions is all too successful. The disastrous effects of chauvinistic propaganda are too obvious to require comment. Activities of reform, as of social delinquents, are in part deliberate attempts at favorable alterations of volitions. True education strives to enhance the powers of making individual decisions freely while propaganda and indoctrination aim to stifle these powers.

Even concerning ourselves, we are willing to admit one degree of predictability. It is pleasing to be described as "steady," "reliable" and "dependable," but not *too much so.* We would abhor complete dependability as something we associate with simple controllable mechanisms, with the doorbell which cannot ring by itself and cannot help but ring when the button is pushed. We are instinctively wary of persons who seem to have a keen faculty of discerning our motives, who can "read" us, a tacit admission that we know our volitions to be somewhat predictable. And our aversion to persuasive "sales talks" certainly implies that we suspect our decisions to be susceptible to manipulation.

Another body of pertinent evidence is the effect of various stimulants, depressants, intoxicants, tranquilizers and other behavior-modifying drugs, many of which produce their effects in a highly predictable manner. From the use of drugs in mental hospitals to make patients more manageable, it is but a step to the situation portrayed in Aldous Huxley's *Brave New World,* a satirical novel concerning a society in which many problems are solved by administering drugs to the whole population.

It might be argued that the behavior of persons under the influence of drugs has no pertinence to the free will problem for normal individuals. Our own bodies are, however, producers of highly powerful drugs, the *hormones,* secreted by the pituitary, thyroid, adrenal and other glands. These, together with various other chemical substances normally present in the human body, regulate our physiological and, to a considerable extent, our psychological condition. Thus, a deficiency of thyroxine, the active hormone of the thyroid gland, may produce a dull, lethargic condition; an excess may cause overactivity and nervous excitability.

It is a moot question as to whether these substances are to be thought of as part of "man's nature" from which his volitions are presumed to arise. If a patient's bodily chemistry departs from the normal so greatly as to produce marked psychiatric symptoms, we are apt to think of him as a victim of alien influences. But even within the normal range, variations in bodily conditions affect mental dispositions in a predictable manner. "Tired and irritable" or "fresh and cheerful" are well-known expressions of such body-mind relations. The wife who postpones the discussion of a new wardrobe until her husband is well fed and relaxed is taking advantage of the fact that volitions may be manipulated predictably by altering the physical condition of the body.

Habit is yet another factor which limits our choices to an extent of which most of us are hardly aware. Many of these are *group* habits, small amenities of social behavior, manners of speech, which we adopt in a barely conscious but compelling desire to "belong" to our particular niche in society. These are the behavior patterns which would serve to differentiate immediately a convention of salesmen or of scientists in a hotel lobby, and which a child is impelled with the greatest urgency to adopt when it is transferred, by a move of the parents, to a new school.

Konrad Lorenz, an internationally renowned naturalist and a keen observer of animal behavior, which he finds highly pertinent to human affairs, recounts in his book *On Aggression* an incident concerning habit in the life of his pet graylag goose. She had the habit of first turning toward the window before mounting the stairs to follow him to bed at night, for no other reason than that she had happened to do so the first time. One evening, there having occurred an unusual delay, the goose, because of the late hour, made straight up the stairs without first going through her usual ritual of turning toward the window. But halfway up she suddenly exhibited signs of great alarm and then, says Lorenz:

> . . . ran hurriedly down the first five steps, and set forth resolutely, like someone on a very important mission, on her original path to the window and back. . . . I could hardly believe my eyes. To me there is no doubt about the interpretation of this occurrence: the habit had become a custom which the goose could not break without being stricken by fear.

We humans, avers Lorenz, are inhibited by personal rituals no less compelling. More important, however, are those which reach back, beyond single lifetimes, to the history of our culture:

> Oaths cannot bind, nor agreements count, if the partners to them do not have in common a basis of ritualized behavior standards at whose infraction they are overcome by the same magic fear as seized my little greylag on the staircase in Altenberg.

Finally, useful insights into the freedom of the will may be obtained by considering the operation of electronic computers. This is not to say that there is a close analogy, as is sometimes alleged, between "electronic brains" and the brain of man. The real brain is, both structurally and in its mode of operation, enormously more complex than any computer which could be envisioned in terms of presently known techniques. In the cerebral cortex alone there are over ten billion (10^{10}) individual nerve cells, intricately interrelated by more than a trillion (10^{12}) connections which differ in structure and action in many subtle ways. In a computer a single operation may be traced readily through a distinct sequence of elements with only a few branchings between storage "memories," input, output and calculating components. Most mental processes are only diffusely localized and involve at every stage many nerve cells and many complex interconnections. Nature frequently accomplishes results in ways which seem unnecessarily and distressingly complex to the engineer.

A computer may be designed and programmed to play a good game of chess. It does this by surveying, after each of the opponent's moves, all possible moves of its own and all of the opponent's possible countermoves for a number of moves ahead. It then makes the one move most advantageous for itself, or one of several equally advantageous ones. This is much the same as the method used by an experienced human chess player. Only limitations of cost and well-nigh impossible complexity make it impracticable to construct a perfect, unbeatable computer, one which would survey all possibilities far beyond the capabilities of any human chess master. With a machine of this capacity it would be quite out of the question, even for its designer, to *predict* the moves it makes. To do so, it would be necessary to go over all the billions of billions of steps which the computer makes to arrive at the best decisions; and this might well require more than one lifetime.

Here then is a device which exhibits, in effect, the attributes of free will to a marked degree. Its activity is in actual practice quite unpredictable and, in effect, freely chosen. Yet every one of the many events occurring within its complex anatomy is in accord with the laws of classical physics and is therefore completely determined.

Chess playing is but one example of the more general enterprise of making decisions as to the most advantageous procedure under a given set of conditions. In this computers have proven highly successful and useful for undertakings in science, technology, business, government and military strategy.

These observations might well be advanced to refute the assertion, often made by the layman and even by some philosophers, that the act of volition, experienced so deeply as undeniably free, cannot therefore be attributed to any cause. Man is indescribably complex in his structure and in his functionings, and this complexity constantly gains accretions from all his acts and all his experiences. His inscrutability to others, and even to *himself*, may well be the source of his freedom.

Various other matters bearing on the freedom of the will could be brought to this discussion, but what has been said suffices to demonstrate that our volitions are not quite as free of causal factors as a naïve intuition might lead us to believe.

FREE WILL AND CAUSALITY

THE Newtonian concept of a universe consisting of hard, indestructible particles acting upon one another by well-determined, calculable forces was made the basis of a thorough and rigid determinism by the French astronomer and mathematician the Marquis Pierre Simon de Laplace (1749-1827), who said:

> We ought then to regard the present state of the universe as the effect of its antecedent state and the cause of the state that is to follow. An intelligence knowing, at any given instant of time, all forces acting in nature, as well as the momentary positions of all

things of which the universe consists, would be able to comprehend the motions of the largest bodies of the world and those of
the smallest atoms in one single formula, provided it were sufficiently powerful to subject all data to analysis; to it, nothing
would be uncertain, both future and past would be present before
its eyes.

This Laplacian world picture is in accord with the scientific
concept that causality resides in functional relations ("to subject
all data to analysis") and that the meaning of determinism is
calculability. However, whether inadvertently or not, Laplace
neglects one important property of a causal world when he speaks
of the intelligence knowing "all the forces acting in a nature" but
knowing these "at one instant of time." Unless it is understood
that these forces remain unchanged in their nature at all times
(unless, for example, like electric charges always repel and do
not at times change capriciously to attraction), his universe is
not calculable.

His whole concept has been criticized because it involves a
fanciful superhuman "intelligence." This criticism is beside the
point, for the intelligence is but a dramatic device. If the universe
does actually run inexorably in a rigidly deterministic manner
through all of past and future time, it does so whether or not an
intelligence knows about it, for in the universe of Laplace this
knowledge exerts no influence. More valid is the criticism that
the Laplacian conception applies to almost any imaginable kind
of universe, including some highly bizarre ones. There might be,
for example, forces present which, though constant in time, vary
in a complex manner for different particles and different distances
between them. This would require that the intelligence be an
exceedingly capable mathematician, but apparently his capabilities
are unlimited.

Furthermore, the Laplacian description of a causal universe is
useless as a criterion for determining *by experiment* whether or
not it actually is causal in any meaningful sense. To do so, it
would be necessary to have at hand a number of identical uni-
verses, all prepared in precisely the same way and then watched
to see whether or not they run alike. Or the one available universe
would need to be so constituted that it could be brought back
repeatedly to precisely the same initial state. But as things are,

no matter how the one available universe develops, it would be judged by the omniscient intelligence to be a causal one.

Because of his great and well-merited renown in the fields of mathematics and astronomy, Laplace's conception of a completely mechanical and deterministic universe was widely discussed and was given more importance than a careful analysis can attribute to it. But so long as human beings were accorded a unique position apart from inanimate matter and from all other living things, mechanistic theories of the universe could have no relevance to the problem of free will.

The French physician and philosopher Julien Offray de La Mettrie (1709-51) was one of the first to make a vigorous attack upon the barrier separating man from the rest of the universe. He had an excellent grasp of comparative anatomy and physiology, of humans and animals, an inquiring mind and keen powers of observation. His deep sympathy for all living things and his bold and fertile imagination led him to formulate highly original ideas, some curious, some far ahead of his time. He objected strenuously to Descartes's assertion that animals are machines, devoid of thought, feeling or emotion.

La Mettrie conceived of the human soul realistically as a composite of mind, character and disposition, and thus not unaffected by the bodily condition. The state of the soul, he maintained, is affected by disease, both physical and mental, by drink and drugs, and even by diet. (He attributed the, to him, degraded character of Englishmen to their diet of rare and bloody meat.) He saw no difference in kind but only in degree between man and the lower animals. To him there was a clear evidence of an essential unity in all of creation.

Only because of pride and prejudice, he maintained, do men refuse to accept the truth: "Matter contains nothing base except in vulgar eyes which do not recognize it in its most splendid works." He earnestly believed that his philosophy would generate wisdom, tenderness and kindness, and reverence for all of nature. "Convinced, in spite of the protests of his vanity, that he is but a machine or an animal, the materialist will not maltreat his kind . . . he will not wish to do to others what he would not wish them to do to him."

While many of La Mettrie's ideas were naïve (he believed that kind and gentle animals are the most intelligent, and was certain that an exceptionally bright young baboon could be taught to speak), many of his conclusions were restated by others, only many years later on the basis of the works of Darwin and later findings in the biological sciences. Most importantly for the free will problem, he brought forward the idea that the processes occurring in living organisms, including man, are governed by the selfsame laws of the physical sciences which are valid in the inanimate world.

This is the essential idea which produced the long controversy between the proponents of materialism and vitalism. By the latter, Laplacian determinism, as applied to human activities, was represented as making of man a senseless automaton in the grip of a relentless demon which dominates his every thought and feeling and action, the ultimate horror being that the automaton is not even aware of his condition.

Free will, ethics, moral responsibility, the justification for praise and blame, for heaven and hell, were thought to stand or fall on the issue of whether or not the laws of physics and chemistry hold in the body and mind of men.

In all of the discussions which this question has fomented, the words of Alfred North Whitehead (1861-1947), the English mathematician, philosopher and logician, stand out for their clarity and good sense. To ask whether the scientific laws of the inanimate world can "explain life," he says, implies that there exists a corpus of such laws, complete and final, to which the phenomena of life can be referred for solution; and this is of course not the case. All attempts to separate the universe sharply into the categories of the living and the nonliving have come to naught. It is, therefore, logically tenable to assume that there is one system of laws for all things. Much of this has come to be known, not from observing parts in isolation, but from studying the new phenomena which *emerge* whenever more complex groups are formed. Thus, no close scrutiny of protons, neutrons and electrons would disclose how they combine to form atoms; and the study of atoms by themselves could never show what a rich variety of molecules they can produce. Similarly, no molecules, even the most complex ones, exhibit the marvelous properties of the living cell, nor could anyone have foretold that combinations of such cells might paint madonnas or write symphonies. If the terms "physics" and "chemistry" are restricted to complexities below those of living things, then the answer to the question is, of course, negative; but such a restriction is arbitrary. From another point of view, says Whitehead, the question concerns the capabilities of the human intelligence, for which it certainly is not possible to set distinct limits.

As with most controversies about which there has been much heat and little light, the contention that the deterministic laws of physical science coerce the will loses all force when the nature of these laws is properly understood. To repeat, even at the risk of tedium, the scientific conception of causality is quite devoid of any idea of compulsion. If there were compulsion of the human will, it could not possibly have its source in the laws of science. To realize how absurd this notion is, one need entertain only for a moment the bizarre idea that the planets are very unhappy at being compelled to move in elliptical orbits when they would much prefer triangular ones. Continuing to speak metaphorically, it is equally tenable to assume that they were delighted with their elliptical orbits and would have no other. These are indeed the only orbits that are in accord with principles of nature; and to the best of our knowledge they are orbits of all planets everywhere throughout the universe. It is no grave violation of the laws of semantics to say that it *lies in the nature* of planets to move in this fashion.

In harmony with all of creation man also acts in accord with his nature. But he alone does so with consciousness of his volitions and awareness of his acts. Furthermore, unlike the planets, he is at no time of his life a finished product. What he is and what he is to become is the result of all his endowments and his experiences, including his volitions and his acts. He is free to will and also free to strive and to grow.

The idea that the will is coerced by the laws of science is, in fact, quite illogical. For coercion implies a *clash* in which a stronger, alien will overcomes our own. Someone whose will is coerced would be *possessed* by a conscious antagonistic spirit forcing him to will what he would not will. The consciousness which is essential to willing is obviously already present in the one who is doing the willing, and it is senseless to postulate a second consciousness dominating this one. It is true that some of our decisions are arrived at after a consciously experienced contest within ourselves of conflicting drives and motives. But we never experienced this as a clash between ourselves and something alien; and the final decision is always, to us, our own. Most assuredly, we never experience a clash between our will and the laws of chemistry and physics.

Far from being incompatible with free will, determinism in the scientific sense is actually a prerequisite for it. Without a uniform and dependable external world in which it can be put into effect, our willing would be futile. Nor could free will exist in any significant and valuable sense without a permanent and consistent basis in our own natures. If the will were free in the sense of being a thing of the moment, of chance and caprice, not related to anything we call our own, our volitions would be startling, unaccountable and senseless, and our acts would be those of a madman, bereft of all reason, purpose or dignity.

Finally, as it applies to human affairs, causality has been represented by some writers as akin to *fatalism*. But fatalism is concerned, not with what men will or do, but with what happens to them, that is, with what is fated to happen to them in spite of anything they might will or do. In a dangerous situation a believer in causality will take precautions to avoid causes of injury; but a fatalist will stride out boldly, convinced that he can do nothing to bring him safety or harm. Fatalism is the belief that all happenings are determined from the beginning of time for all eternity, that men are mere pawns in the hands of a supernatural power and have no control over their destiny. This is set forth in the words of the Persian astronomer, mathematician, poet and philosopher Omar Khayyam (1050-1123) (as translated by Edward Fitzgerald):

> With Earth's first clay they did the last Man knead,
> And there of the last Harvest sowed the seed:
> And the first morning of creation wrote
> What the last dawn of reckoning shall read.

· · ·

> The moving finger writes; and having writ,
> Moves on; nor all your piety nor wit
> Shall lure it back to cancel half a line,
> Nor all your tears wash out a word of it.

ETHICAL PROBLEMS

OF all the aspects of the free will controversy those relating to *ethics* have always been the most poignant. Ethical problems arise from considerations which have already been discussed. Moral responsibility is thought to require free, uncaused volition; and Laplacian determinism, carried over into human affairs by the accomplishments of nineteenth-century biological science, is interpreted as denying that freedom.

Some philosophers, convinced that this contradiction cannot possibly be resolved, seek to avoid it by placing science and ethics in two unrelated categories, and establishing free will as an unassailable truth on grounds of ethical laws and the dictates of intuitive psychology. The French philosopher Auguste Valensin in an article published in the *Etudes Philosophiques* of 1953 writes:

> Two classical demonstrations establish, according to our opinion, in a satisfactory way, the existence of free volition; the first one is moral and has to induce us to believe in liberty; the second one is psychological and has to confirm this belief. . . . Does not everyone know by experience what it means to assume the moral responsibility of an action? To assume such responsibility is identical with perceiving that the will is free.

In this simple and direct statement there is no indication that the will is related to a cause of any kind. It is, in fact, just in ethical matters where the layman is apt to be most insistent that the will must be quite spontaneous. A man, he will say, cannot be held responsible for what he is *made* to do, and it matters not whether the compulsion comes from outside influences or from within himself. The essential condition for free will is held to be that a man *could have chosen to do otherwise than he did do.* But if, it is argued, a given man in a precisely given state could freely make any one of several quite different choices, there could be nothing within him causing him to make the one choice he actually did make.

This seemingly plausible argument does not stand up under inspection. The fact is that the man made just this one choice, and it is impossible to give him a second chance under identical conditions both around him and within him. His own subjective conviction that he could have made a different choice the first time proves nothing except perhaps that he has a fertile imagination.

A similar argument has been advanced with regard to *remorse*. The acute feeling one has concerning a regrettable act, that one could have and therefore should have acted otherwise, is advanced to prove that one and the same person could at a given moment have acted in any one of several significantly different ways. This is again an illusion based on the erroneous idea that a man is a fixed and unchangeable entity. The remorseful person is, however, "sadder but wiser," due in part to what he has experienced because of his decision. At the time he is feeling remorseful, he is no longer quite the same person he was when he did what he now regrets.

There are further considerations in the ethical realm which point to a causal relation between a man's volition and something permanent in his character. Bestowing praise or blame on a person for his acts would not be justified if they were not in a real sense his own. And punishment can be effective only if it contributes to an improvement in character. The work of a lazy pupil can be improved by a spanking, by adding a factor to his environment which tends to make him more industrious. But spanking a boy who fails in his work because of stupidity is ineffective because it cannot change the element in his nature which is responsible for his poor performance.

Ethical judgments are, in general, not about acts or volitions but about people. We do not judge an act to be good; we judge the one who performs it to be a good person. When we say that someone is "kind," "generous" and "brave," or "mean" and "cowardly," we are describing character. Acts which are clearly unintentional are deemed neither blamable nor praiseworthy. We do not blame someone for an unfortunate accident unless we lay it to his innate carelessness, again a trait of character. Self-blame, as when we reproach ourselves for being careless or inconsiderate, is a judgment of our own character. We feel less regret for acts which seem to be due to external factors not closely related to our own selves. All of these common observations serve to reinforce the conclusion that in ethical matters our volitions are not uncaused but are closely related to our character.

The contemporary philosopher C. A. Campbell contends, however, that in some instances our volitions may be quite spontaneous and therefore not compatible with any principle of causality. His argument may be exemplified by the action of a thoroughly cautious and conservative man who discovers that a fire is spreading through his neighbor's house, in which a small child is sleeping unattended. It lies in his *character* as a dutiful citizen to call the firemen promptly and inform them of the child's peril. Realizing that this help may come too late, he rises heroically to the occasion and saves the child, in an act apparently quite contrary to his well-formed character.

Campbell believes that most of us are capable of such spontaneous *creative* volitions, that they are genuinely uncaused and are yet of great moment for the ethical value of the free will. While this conclusion is no doubt worthy of consideration, it disregards the fact that all responses of all living things are conditioned by the nature of the stimulus as well as that of the organism upon which it acts. A highly unusual stimulus might elicit an equally unusual response. Furthermore, while innate character is normally rather stable, it might change rapidly and drastically under great stress. This supposition would be substantiated if it were found, as is entirely possible, that the character of our cautious individual had undergone a notable and permanent change because of this episode.

From the whole of these discussions concerning causality and the nature of the will, it is evident that the common, intense, subjective conviction that our volitions are freely our own, that we are not ridden by an alien potency which could make us will what we would not will, arises simply from the fact that our volitions, though they are determined in part by the stimuli of our total environment, are in harmony with our own natures. This conclusion is in accord with the preponderance of philosophical opinions and with the principles of classical science.

18

Determinism, Free Will and the New Physics

WHEN, in the early 1930's, Professor Heisenberg came to the United States to give a series of lectures on quantum mechanics at the Massachusetts Institute of Technology, his fame as the originator of the uncertainty principle had preceded him. The audience at his first lecture filled to overflowing the largest lecture hall at the Institute, for he was hailed as the champion who had released the human will from the bondage of Laplacian determinism.

The argument to refute determinism on the basis of the uncertainty principle goes as follows: Laplace asserts that a superhuman intelligence can know all of the past and future by making *calculations* based upon a *complete and ideally precise knowledge* of the *positions and motions* of all the particles in the universe. The uncertainty principle shows, however, that a precise knowledge, attained simultaneously, of both position and motion (momentum) of particles is impossible, even in principle. Thus, at one stroke, Laplacian determinism is destroyed.

This conclusion, however, does not follow quite so simply and directly. In the first place, the uncertainty principle alters things significantly only in the tiny realm of atomic phenomena. For everyday affairs, those which are of real concern in our lives, this uncertainty, as measured by the magnitude of Planck's constant, is far below the limits of observation, far too insignificant to afford any comfort to the freedom of the will. If this constant h were numerically greater, by a factor of many billions of billions, quantum mechanical uncertainty *would* obtrude directly into our lives. But, as has already been made clear (page 102), the universe would then be such that neither volition nor any other mental activity would exist.

A more searching analysis is given by Professor Henry Margenau, who, it will be recalled, opts for the reality of the de Broglie matter waves rather than for the particles. He contends that quantum mechanics has by no means abolished determinism. On the contrary, says Margenau, seeing that some of the new experimental findings in atomic physics threatened classical determinism, physicists developed quantum mechanics to *reestablish* determinism on a new basis although, he admits, some did not realize that this was what they were doing. He sees the relation of classical mechanics and quantum mechanics, not as a matter of determinism versus indeterminism, but as a difference in the way in which the *state* of a physical system is described. Classical mechanics does this in terms of position and momentum of the particles while quantum mechanics uses ψ-functions, patterns of matter waves. In terms of these new "state variables" quantum mechanics is completely deterministic.

With regard to the Laplacian intelligence, says Margenau, its calculations must necessarily deal with the mathematically predictable course of the matter waves. All of what the intelligence can know about the universe, even in principle, is thus just as rigidly determined as before; and nothing is changed.

Margenau's basic premise, that reality resides in the waves rather than in the particles, may be challenged. But his conclusion concerning causality holds as well in a world of particles. The fallacy in the argument against Laplace is that it envisions particles in both of two incompatible ways. It presents them first as classical Newtonian particles, which are so readily envisioned intuitively, and then frees them of determinism through the uncertainty principle so that they can, as it were, do individually as they please. Actually, the uncertainty principle speaks of enti-

ties, very different from classical particles and also very difficult
to grasp intuitively, for which simultaneous precise position and
momentum *does not exist*. One cannot have it both ways. Either
one must stay with the classical particles, whose individual behavior is thoroughly deterministic, or one must have the quantum
mechanical "particles," for which only deterministic statistical
laws, represented by wave patterns, have meaning. In either case
the proper conclusion is that the intelligence can know only that
which is causally determined.

Nevertheless, this very characteristic of quantum mechanics,
that its determinism with regard to the activity of particles is statistical, has been seized upon by those who would free the will.
Statistical laws are represented as mathematical and therefore
mental in nature and subject to the influence of "mental forces."
This idea may be countered by emphasizing the very nature of
quantum-mechanical indeterminism: individual events at the
atomic level are not merely undetermined; they cannot possibly
be determined by anything, including mental forces. To be effective, an influence must be capable of altering the statistical picture; and quantum mechanics shows that this requires good old-
fashioned classical forces.

In still another essay at rescuing free will, it is alleged that
mental activity might originate from the action of a single atom
or even a single electron in the brain and that the uncertainty
principle could therefore introduce an element of pure chance
into mental events and thus really free the will. Here it is useful
to consider again the chess-playing computer. Pure chance would
be introduced into its operation if it had several faulty, loose
electrical connections which rattle and make intermittent contacts
with random vibrations of the building due to steps of the per-
sonnel, passing traffic or perchance an occasional slight earth-
quake. Such a computer could not possibly play a sensible game
of chess nor could it make any other sensible decisions. This is
merely a special instance of the general conclusion, arrived at
earlier, that volitions based upon chance and caprice are those
of an idiot.

Furthermore, it is a well-established basic law of neuro-physi-
ology that a nerve cell will not respond until the stimulus
reaches a certain threshold value, whereupon it responds with

maximum strength (the law of "all-or-none" response). Any event so minute as to be sensibly effected by the uncertainty of quantum mechanics would certainly lie far below the threshold for the initiation of any kind of brain activity.

The idea that the new physics is not deterministic and therefore somehow "spiritual" seems quite persistent. Concerning the jumps of individual electrons between the quantized orbits of the Bohr atom model, Bernard Bavink, who has been quoted earlier, writes:

> We must remember firstly, that the individual elementary act (of jumping) as such is not calculable, but left free; secondly, that the real essence of this freedom is perhaps or probably a psychical event.

In spite of the cautious "perhaps or probably," Bavink does not hesitate to draw the conclusion that man's will certainly cannot be less free than a single electron. This passage has the virtue of pointing out that the indeterminism which came to the fore with de Broglie's matter waves had already appeared a decade earlier in Bohr's concept of electron jumps, where it had been ignored by physicists, probably in the hope that further enlightenment would make it go away.

The electron in the Bohr atom model is not quite as free as Bavink assumes. Each downward electron jump is related to the emission of a photon of one particular wavelength or color; and the statistical distribution, for the jumps of the electrons in all of the atoms of a given light source, must be in accord with the predictions of the quantum-mechanical laws governing the relative intensities of the various colors in the whole emission spectrum. If each electron were free to jump as it pleases, the total effect would be chaotic. In order to achieve the proper intensity relations in the emitted light, the electrons, to continue Bavink's metaphor, would have to agree somehow on appropriate joint action. What is forgotten in these attempts to achieve freedom in the new physics is that it has statistical laws which are not equivalent to caprice. In a crude analogy, these laws may be likened to the parking laws of a town. These do not prescribe a definite position for each car; nevertheless, they do determine the overall pattern in which the cars are arranged.

Bavink is not alone in his opinion that the new physics has freed the will. *In The Mysterious Universe* Sir James Jeans writes:

> . . . the picture of the universe presented by the new physics contains more room than did the old mechanical picture for life and consciousness to exist within the picture itself, together with the attributes which we commonly associate with them, such as a free will.

This is possible, says Jeans, because the determinism of quantum mechanics "is a determinism of waves and so, in the last resort, of knowledge." A determinism "of knowledge" makes sense if it is taken to mean that deterministic *waves* yield *knowledge* about the course of events. But Jeans proceeds to draw the further conclusion: "Things still change solely as they are compelled, but it no longer seems impossible that part of the compulsion may originate in our minds." The use of the word "compulsion" by a scientist with regard to the course of events is certainly surprising. But even more so is the implication that the new physics permits our minds to exert this compulsion. This is again the fallacy concerning the power of "mental forces."

The English astronomer Sir Arthur Eddington (1882-1944) is even more explicit concerning the liberation of the will by the new physics. He believes firmly that classical determinism did actually inhibit the freedom of the will; in fact, he makes it almost synonymous with fatalism. In *New Pathways in Science* he asks what it would avail him to struggle to give up smoking if "the laws which govern the matter of the physical universe already preordain for the morrow a configuration of matter consisting of pipe, tobacco, and smoke connected with my lips?"

Having thus established the need for rescuing the will, he asserts that the new physics has accomplished this task. In the essay "Science and Religion" he writes:

> I think there is no longer any need to doubt our intuition of free will. Our minds are not merely registering a predetermined sequence of thoughts and decisions. Our purposes, our volitions are genuine; and ours is the responsibility for what ensues from them.

That Eddington attributes this genuine freedom of the will, and the consequent moral responsibility, to the new physics is brought out very clearly by statements such as: "I attach importance to the physical unpredictability of my pen because it leaves it free to respond to the thought evoked in my brain . . . ," a naïve assertion that "physical unpredictability" permits his free volition to guide his pen.

He does, however, reject the idea that the unpredictability, according to the uncertainty principle, of the single atom's activity could be significant for mental events. "Could we," he asks, "pick out one atom in Einstein's brain and say that if it had made the wrong quantum jump there would have been a corresponding flaw in the theory of relativity?" Rather he believes that "we must attribute to the mind power to affect systematically large groups of atoms, in fact to tamper with the odds on atomic behavior." Eddington seems to believe that quantum mechanics gives the mind the power to alter the statistical predictions ("tamper with the odds") of quantum-mechanical laws. As has already been noted, however, the principles of quantum mechanics show quite definitely that these statistical laws are completely deterministic and that the statistical pictures, "the wave pattern," is altered, not by mysterious mental influences, but by real physical forces.

Clearly, the opinions of Jeans and Eddington, and of many others who wrote in a like vein, are based upon serious misapprehensions concerning the spirit both of the new physics and of the classical science which preceded it. Yet their reactions to the impressive new scientific conceptions of our century are understandable. These two scientists, and others of their generation, had their roots in the late nineteenth century. In their youth the problems of free will, and more generally of ethical and spiritual values seemingly threatened by deterministic science, were still very much a topic of serious discussion by thoughtful people. It is not surprising that they were much taken by developments in science itself which seemed to offer an escape from these difficulties.

That so many persons with an interest in philosophical matters flocked to hear Professor Heisenberg discuss the uncertainty principle is proof enough that misconceptions may become widespread and may have a powerful and persistent influence on men's thoughts.

Thus, it is fortunate that the human will stood in no need of being rescued from bondage by the new physics. As previous discussions have made quite evident, the conception of causality propounded by scientists throughout the classical period poses no threat to the freedom of the will.

DIVERSE INTERPRETATIONS

THE writings of Max Planck concerning free will and determinism in the new physics, to which he contributed so importantly, are well worth discussing as a study of the way in which an attachment to the ideal of classical causality and an emotional aversion to the metaphysics of quantum mechanics could color the views of a scientist inclined to philosophical speculations. His ideas are set forth in a small book, *The Philosophy of Physics*, from which the following quotations are taken.

Planck conceives of causality as a basic and indispensable principle, not only for science, but for the rational and meaningful conduct of life. It "cannot be demonstrated any more than it can be logically refuted: it is neither correct or incorrect, it is a heuristic principle . . ." That is, it is essential to learning and discovery. "The law of causality lays hold of the awakening soul of the child and compels it to continually ask why." To him causality is inherent in the nature of things. It is unthinkable that it should be contingent upon experimental findings and that it should have been set aside by recent observations in atomic physics.

In support of his views he points out that science has always worked in two different worlds. One is the "world of senses," of observation and experiment. Here it has never been possible to demonstrate causality conclusively; because of unavoidable errors of observation two like experiments never show exactly the same result. Causality exists strictly only in the "physical world image," a purely intellectual world of ideal mathematical laws which is abstracted from the real world of the senses. Scientists deal with observations by transferring them from the world of the senses to the world image, analyzing them there by means of deterministic laws and then transferring the results back to the world of senses. The uncertainties of observation are, in effect, eliminated by this double translation between the two worlds. This, says Planck, has always been the situation in classical physics.

He then asserts that in the new physics this same situation still obtains. There are the quantum-mechanical laws of Schrödinger, Heisenberg and Dirac, which portray "fully as rigid a determinism in the world image of quantum physics as in that of classical physics. The only difference is that different symbols are employed and different rules of operation obtain." This is quite true. He then adds, however, that, just as in the case of classical science, the uncertainties in the world of the senses are effectively eliminated by the double translation; but he cautions that it is necessary to realize "how much more difficult it is in quantum physics to translate an event from the world image into the sense world and vice versa." Planck implies that the new physics has changed nothing in either of his two worlds but has merely complicated the process of making the translation between them; that is, it has not indicated a change in the nature of things but only in the way in which scientists deal with them. In particular, the principle of causality remains secure.

Planck then departs completely from physics, both old and new, and in a manner strongly reminiscent of Laplace, proposes to base causality upon an "ideal spirit" whose existence, he says, is implied by "a premonition of a certain harmony between the outer world and the human spirit." He continues:

> The most perfect harmony and consequently the strictest causality in any case culminates in the assumption that there is an ideal spirit having full knowledge of the action of the natural forces as well as of the events in the intellectual life of men; a knowledge extending to every detail and embracing past, present and future.

Thus, Planck goes a long way beyond Laplace by extending the knowledge of the ideal spirit over "the intellectual life of men."

Then he comes to the problem of free will and considers the import of the omniscient spirit. "It may be asked," he says, "what becomes of human free will on this assumption, and it may be suspected that by it man is degraded to the rank of a mere automaton." To this he replies:

> In my opinion there is not the slightest contradiction between the domination of a strict causality in the sense here adopted and the freedom of the will. . . . The notion of human free will can mean only that the individual feels himself to be free, and whether he does so in fact can be known only to himself. Such a state of affairs is entirely compatible with the fact that his motives could be apprehended in every detail by an ideal spirit. A feeling that such a state of affairs is derogatory to the ethical dignity of the individual implies an obliviousness of the vast difference between the ideal spirit and the intelligence of the individual.

But even this sublimated causality does not suffice to establish for man the freedom of the will. "For this purpose," says Planck, "he must refer to a totally different law, namely, the law of ethics, which is based on a different foundation and cannot be comprehended by scientific methods." Thus, Planck comes finally to the support of free will by the imperative of ethical duty. This he sums up in the dictum: "The law of causation is the guiding rule of science; but the Categorical Imperative—that is to say, the dictate of duty—is the guiding rule of life."

That two thoughtful scientists, Eddington and Planck, could come to diametrically opposite views on the basis of the same factual considerations affords an interesting insight into the relation of science and philosophy, into the ways in which contemplation of science can influence but cannot determine philosophical conclusions. Eddington is certain that classical science was inimical to the freedom of the human will and is just as certain that the new physics has abolished causality and set the will free. Planck is convinced that causality still exists much as before but that the will is nevertheless free on grounds entirely apart from science.

CRITICISMS

THE concept of causality which came to the fore in the seventeenth century with the birth of classical science is concerned, not with ephemeral and contingent events, but with the universal and enduring relations among these events. The new physics also has its causal laws. These do not refer, however, to relations between physical events but between mathematical abstractions which pertain to events only as probabilities. This most recent kind of scientific causality is a sophisticated one, not easily reconciled with older concepts, either in philosophy or in science. The objection has indeed been raised that probability or *chance* is the very antithesis of causality and that their juxtaposition in quantum mechanics is contradictory.

The new physics has occasioned other metaphysical difficulties. Opinions have been voiced both by philosophers and by scientists that a theoretical system which fails to account for individual happenings in detail is essentially tentative and incomplete. There have been objections to the dual concept of waves and particles and to the strange nature of some entities in quantum mechanics which lack the very attributes deemed most essential to the intuitive idea of reality. Further, it has been thought intolerable to set arbitrary limits to the precision with which the state of things might be determined.

One of the most ambitious attacks on quantum mechanics is that of the American physicist and philosopher David Bohm, set forth in his book *Causality and Chance in Modern Physics* (1957). This is an elaboration of ideas advanced some two decades earlier by de Broglie himself, ideas which he abandoned but was moved to reconsider with the aid of a young colleague, Jean-Pierre Vigier, when the work of Bohm was published. He attempts nothing less than a reformulation of quantum mechanics proposing to achieve precision beyond the limits of Heisenberg's uncertainty principle and to reestablish a kind of causality which is related directly to actual events among real physical entities.

To indicate how this might be realized, Bohm cites an analogy in nineteenth-century physics. The analysis of the properties of gases, conceived as swarms of molecules, required ideas of probability and statistical methods. But at a more fundamental level the statistical laws were based on the strictly deterministic laws of classical mechanics pertaining to the motions of the individual molecules. Similarly, he maintains, it is possible to relate the statistical laws of quantum mechanics to certain "hidden variables" obeying deterministic laws at a deeper level.

Bohm and his collaborators thought for a time they were on the verge of achieving important successes in those areas, such as very-high-energy particle phenomena, in which quantum mechanics has been encountering difficulties. But in the following years nothing has come of this, nor have any concrete ideas been advanced concerning the nature of the "hidden variables" or the laws which supposedly determine their actions.

Because of this lack of tangible results, either in furthering theoretical research or in accounting for experimental findings, Bohm's work has been received with much reserve by the majority of physicists, a circumstance which has given rise to charges of dogmatism and of unjust opposition to new ideas. This situation is summed up nicely by the English philosopher of science N. R. Hanson in a quotation from an essay entitled "Quanta and Reality," prepared for presentation in 1961 on a distinguished BBC Third Program broadcast:

> So it is reasonable for Bohm, Vigier and Feyerabend to continue to speculate—and unreasonable for any practising physicist to dissuade them from doing so on the quite specious grounds that quantum mechanics is a beautiful success as it now stands. But it is also reasonable for practising physicists to act on the assumption that Bohm, Vigier and Feyerabend will not succeed with the development of their speculations, and unreasonable for anyone to interpret this lack of confidence as a symptom of dogmatic orthodoxy.

The root of all these contentions is in large part a deep human predilection for a direct causality in the affairs of the universe, a conviction that a precise causality of individual events is "of course" to be preferred. Yet this kind of causality was hardly tenable even before the coming of the new physics. When examined critically, the deterministic laws of classical science are seen to be idealizations of actual experience; for all our observations are necessarily concerned with objects occupying portions of space and events occurring in portions of time, portions which, no matter how minute we succeed in making them, yet include a *finite* range. Classical laws deal, however, with *points* of no extension in space and *instants* of no duration in time.

The concept of space and time which seems to be entertained by those who hope for a return of classical causality is much akin to that proposed in the seventeenth century by Sir Isaac Newton. Space and time are thought of as a grand stage upon which objects are arrayed and events occur, a stage which has independent existence even when empty and dark. Nearer to actuality is the space and time of the psychologists, who point out that we are never aware of space itself but only of the spatial relations of objects, nor can we experience time except through the temporal sequence of events. An ideally precise causality requires the ideally precise location of points in space and of instants in time, in a smooth and continuous kind of space and time whose existence has never been demonstrated.

All this was known to thoughtful scientists long before the coming of the new physics. Quantum mechanics and the Heisenberg uncertainty principle have not "destroyed" causality. These advances may well be a step toward making its meaning clear.

Here the story rests, but it is by no means ended. What must have emerged most clearly is that much of importance remains to be explored. Some of the discussions of what happens in our universe have no doubt had the feeling of pertaining to matters rather well explored and substantiated. Others, perhaps the more interesting, were obviously tentative and subject to surprising revisions. The excursions into philosophy had the burden of showing that, in its search for propositions which are timeless, philosophy may well contemplate science with detachment but cannot ignore its findings. And as science strives toward an ever more comprehensive description of nature, it must look to philosophy for guidance in formulating its tenets. Yet each has its own ways; neither is suited to cope with the problems of the other.

Glossary of Scientific Terms

For Additional Reading

Index

Glossary of
Scientific Terms

Note: Additional discussion of many terms may be found by consulting the Index.

ACCELERATION, LINEAR: Time rate of change of velocity along a line. A commonly used unit of acceleration is a rate of change of one centimeter per second in one second.

ACTION: The product of energy and time, expressed in erg-seconds.

ALL-OR-NONE LAW, of nerve action: A single nerve fiber does not respond until the stimulus attains a certain threshold value, whereupon the response is maximum and does not increase for a stronger stimulus (which may, however, cause a greater number of nerve fibers to respond and to do so in more rapid succession).

AMPERE: See ELECTRICAL QUANTITIES AND UNITS.

AMPLITUDE: The maximum displacement from the central position for an oscillating motion or a wave.

ANTIMATTER: Matter consisting of antiatoms made of antiparticles.

ANTIPARTICLE: A particle identical in mass and in various other properties to the corresponding ordinary particle but having an electrical charge and certain other characteristics of opposite sign. When a particle and its antiparticle collide, they are annihilated, with the creation of photons or mesons.

ATOM: The smallest particle of a chemical element that can exist either by itself or in combination with other atoms to form a molecule.

ATOMIC MASS UNIT (AMU): The unit in terms of which the relative atomic weights of the chemical elements are expressed, the atomic weight of the most abundant isotope of oxygen being defined as exactly 16 AMU. One AMU equals 1.661×10^{-24} gram, equivalent to 931 Mev of energy.

ATOMIC NUMBER: The number designating the position of a chemical element in the periodic table. The atomic number of each element is equal to the number of electrons in an atom of that element.

BARYON: Any particle having a half-integral spin value and a mass equal to, or greater than, that of the proton. Baryons having a half-life less than 10^{-22} seconds are usually called *resonances*.

BARYON NUMBER: A number which has the value $+1$ for all baryons and -1 for all antibaryons. It is conserved in all particle interactions: the sum of the baryon numbers of all the particles entering an interaction is equal to that for all the particles resulting from the interaction.

BILLION-ELECTRON-VOLTS (Bev): A unit of energy, and of equivalent mass, equal to 10^9 electron-volts. (Note: The American billion is one thousand million (10^9) while the British billion is one million million (10^{12}). To avoid this ambiguity, the term *Giga-electron-volts* (Gev) is frequently used in the scientific literature for 10^9 electron-volts.)

BOHR UNIT: A unit of angular momentum or spin equal to $h/2\pi$ erg-seconds.

BOOTSTRAP DYNAMICS: A theory of particles and their interactions based on the assumption that all the baryons and mesons, together with their modes of interaction, form a closed, self-determining system.

CEREBRAL CORTEX: The upper part of the brain in which most of the conscious nerve activities related to sense perception, conscious muscular control, reasoning, emotions, and so on, are located.

CHARGE MULTIPLET: A group of particles which differ in electron charge and vary slightly in mass but are identical in many other respects.

CHRONON: A hypothetical ultimate unit or briefest possible interval of time, usually estimated to be about 10^{-24} second.

COMPOUND, CHEMICAL: A chemical substance whose molecules are composed of two or more kinds of atoms. The chemical elements forming a compound are always present in definite proportions by weight.

CONSERVATION LAW: A law expressing the conservation or constancy of some quantity during a physical or chemical change. Example: the conservation of electrical charge during particle interactions.

COSMIC RAYS: High-enery particles, mostly protons with a small fraction of heavier atomic nuclei, streaming into the earth's atmosphere from outer space. These *primary* cosmic rays, when colliding with air molecules, create *secondary* cosmic rays, consisting of a variety of particles, including high-energy photons and unstable particles.

COULOMB: See ELECTRICAL QUANTITIES AND UNITS.

DIFFRACTION: The bending aside from a straight course of waves traveling in a homogeneous medium when they are passing opaque objects.

DIRECTED QUANTITY: A quantity, such as force or velocity, which possesses both magnitude and direction. The direction of angular velocity and of angular momentum or spin lies in the axis of rotation. The technical term is *vector quantity*. By contrast, a quantity such as mass, which possesses magnitude only, is a *scalar* quantity.

DYNE: See FORCE.

ELECTRICAL CHARGE: A property of many kinds of particles which produces electric and magnetic fields in the surrounding space. Large pieces of matter normally contain equal numbers of positively and negatively charged particles which neutralize each other's electrical effects.

ELECTRIC FIELD: See ELECTRICAL QUANTITIES AND UNITS.

ELECTRICAL QUANTITIES AND UNITS: The basic electrical quantity is the *electrostatic unit of charge* (esu), defined thus: two esu one centimeter apart exert forces of one dyne upon each other. The existence of electric charge creates an *electromagnetic field* (a combination of electric and magnetic fields) in the surrounding space. Charges *at rest* are surrounded by an *electrostatic* field; when they are in *motion*, they produce, in addition, a *magnetic* field. However, an electric current in a wire (a stream of moving electrons) produces only a magnetic field because the electrical effect is neutralized by an equal amount of stationary positive charge in the copper atoms of the wire. An electric field acts on charges, either at rest or in motion, so as to do work on them and change their kinetic energy. A magnetic field produces a force only on *moving* charges, a force which is always at right angles to the direction of motion. It changes the direction but not the magnitude of the motion; it does no work and leaves the kinetic energy unchanged. *Oscillating* electrical charges produce traveling *electromagnetic waves*, which carry energy. (The modern concept assigns this energy to a stream of photons.) The *practical* electrical units are the coulomb, ampere, volt, watt and ohm. The *coulomb* is a unit of electric charge equal to 3×10^9 esu. The *ampere* is a unit of electric current equal to a flow of one coulomb per second. The *volt* is a unit of difference of electrical potential: one joule of energy is required to move one coulomb of charge from a given point to another which is one volt higher in electrical potential. The *watt* is a unit of power (rate of doing work) of one joule per second. The *ohm* is a unit of electrical resistance such that a difference of potential of one volt will maintain a current of one ampere through it. (See also ENERGY; FIELD.)

ELECTRON: One of the very few kinds of stable particles and a constituent of all atoms. It has a rest mass of 9.11×10^{-28} gram, a charge of -4.80×10^{-10} esu (designated as e) and a half-integral spin.

ELECTRON-VOLT (ev): A unit of energy and of equivalent mass. The energy acquired by a freely moving particle having a charge equal to one electronic unit (e) when passing through a difference of potential of one volt.

ELECTRONIC OSCILLATOR: A hypothetical, elastically bound electron in an atom, capable of performing oscillations when the atom is jolted or when an oscillating electromagnetic field passes over it.

ELEMENT, ARISTOTELIAN: Any one of the four assumed basic substances, earth, water, air and fire, of which all material things were thought to be constituted.

ELEMENT, CHEMICAL: Any one of the ninety-two basic substances which cannot be analyzed into simpler components by chemical means and which can combine chemically to form innumerable chemical compounds. The molecules of the elements consist usually of one or two atoms of the same kind. The elements have been arranged, according to atomic weight and chemical properties, in an array of rows and columns called the periodic table of elements. (Compare COMPOUND, CHEMICAL.)

ELEMENTARY PARTICLE: A presumably ultimate element of matter or of radiant energy, such as an electron, proton or photon.

ELLIPSE: A plane oval figure, the section of a circular cylinder cut at an angle to the axis. (When cut at right angles to the axis, the cross section is a circle, which may be thought of as a special case of the ellipse.)

ENERGY: Capability of doing work or, in general, of producing observable effects. No agency can produce any effect on any system unless it is capable of imparting energy to that system. A system loses energy when it does work and gains energy when work is done on it; thus, *work* is energy in transit from one system to another. A force does work when it acts on a moving body in the direction of the motion. The basic unit of energy and work is the *erg;* it is the work done by a force of one dyne acting on an object as it moves through a displacement of one centimeter. The *joule* is a unit equal to 10^7 ergs. *Kinetic* energy is the energy inherent in a moving mass. *Potential* energy is the energy possessed by an object situated in a field of force, as by a mass in a gravitational field. As the mass moves downward, it can do work because of the force it can exert, and while performing work, its potential energy decreases. Similarly, charges in electric fields have electrical potential energy. *Radiant* energy is the energy of photons. *Heat* is the energy inherent in the random motion of the molecules of matter.

ENERGY-MASS RELATION: A relation first proposed in Einstein's special theory of relativity. According to the relation $E = mc^2$, energy and mass are interconvertible at the rate of 9×10^{20} ergs for one gram. (The symbol $c =$ velocity of light in free space $= 3 \times 10^{10}$ centimeters per second.)

EPISTEMOLOGY: The branch of metaphysics which deals with the nature of knowledge, with methods of acquiring and evaluating knowledge and with the relation of the human mind to the external world of objects.

ETHER: A hypothetical invisible, intangible entity thought to pervade all of space and to be capable of transmitting electrical, magnetic, optical and gravitational effects.

FIELD: A region of space in which objects have forces acting on them, such as the space around the earth where the gravitational field produces forces on objects proportional to their mass. Similarly, electrical fields exert forces on objects proportional to their electrical charge. A field may be visualized in terms of imaginary lines of force which point everywhere in the direction of the force and whose density (closeness) is a measure of the strength of the field.

FORCE: A push or pull. It is measured by the magnitude of acceleration it imparts to a mass. The *dyne* is a unit of force which, when acting on one gram of mass, produces a change of velocity of one centimeter (cm) per second in one second (an acceleration of 1 cm/sec²). The *newton* is a unit equal to 10^5 dynes. The gravitational pull of the earth on one gram of mass is 981 dynes. In particle physics the concept of force has been enlarged to include influences which produce various other effects in addition to acceleration, such as creation and annihilation of particles. The four known kinds of forces, in order of increasing strength, are the gravitational, weak, electromagnetic and strong force.

FREQUENCY: The time rate of repetition of any periodically repeating phenomenon, such as an oscillation or a wave. It is usually expressed as the number of oscillations or cycles per second.

GAS: A tenuous form of matter in which the molecules are usually separated widely and move freely between collisions.

GRAVITATION: The universal property of all material objects to attract each other. The attractive force decreases with increasing separation. For spherically symmetrical objects the force varies directly as the product of their masses, m_1 and m_2, and inversely as the square of the distance r between their centers, according to the formula $F = Gm_1m_2/r^2$, where G is the universal constant of gravitation.

GRAVITON: A hypothetical particle which, when passing to and fro in virtual form between two masses, in large numbers, is thought to mediate the gravitational attraction between them.

HADON: A general name for all particles (baryons and mesons) which interact by means of the strong force.

HALF-LIFE: The period of time required for half the atoms in a given sample of radioactive substance to undergo radioactive decay.

HYPERCHARGE: Twice the average charge of a charge multiplet. Example: the average charge of proton-neutron multiplet is $(1 + 0)/2 = \frac{1}{2}$; the hypercharge is 1.

HYPERON: A term formerly used for baryons heavier than the neutron.

IDEALISM: Any one of various related philosophical theories which hold that perceived objects exist only in the mind and that it is impossible to prove that there is an objective "real" world external to the mind. Also known as *mentalism,* it is opposed to *realism,* the doctrine that the world of physical objects exists independently of being thought of or observed.

INORGANIC COMPOUND: A chemical compound characteristic of non-living substances, mostly consisting of simple molecules built of a small number of atoms. (Compare ORGANIC COMPOUND.)

INTENSITY, of a beam of radiant energy: The rate of energy transport by a beam. It is proportional to the energy content of the individual photons and to the number of them passing a cross section of the beam per second. In the wave concept it is proportional to the square of the wave amplitude.

INTERFERENCE OF WAVES: The effect of the superposition of two or more waves passing a given point together. If two waves of the same wave length are in phase (in step) at the point, they reinforce each other (constructive interference); if they meet out of phase, for example, with oscillations in opposed directions, they neutralize each other wholly or partially (destructive interference).

INVARIANCE: The property of remaining fixed under changing conditions. Example: the invariance of shape of a rigid object under change of location.

INVOLUNTARY NERVOUS SYSTEM: The system of nerves which is concerned with activities in the body, such as the processes of digestion, which are not under conscious voluntary control. (Compare VOLUNTARY NERVOUS SYSTEM.)

IRREVERSIBLE PROCESS; A process, such as the intermixing of two initially separate kinds of gases by diffusion, whose spontaneous reversal is prohibitively improbable.

ISOTOPE: Any one of two or more atoms of a given element which have the same number of protons but different numbers of neutrons in their nuclei. Since the isotopes of a given element also have the same number of electrons, arranged in the same way about their nuclei, they have the same chemical properties. Example: Cl^{35} with 17 protons, 18 neutrons; Cl^{37} with 17 protons, 20 neutrons.

ISOTOPIC SPIN: A property of mesons and baryons whose various orientations in an abstract mathematical space distinguish the various members of a charge multiplet from one another. It is significant for the occurrence of particle interactions and for their organization into groups.

KINETIC THEORY: A theory of gases based on the concept of a gas as a swarm of molecules in random motion.

LEPTON: Any one of a small group of particles having half-integral spins and masses less than any of the mesons.

LEPTON NUMBER: A number having the value $+1$ for leptons and -1 for antileptons. Lepton number is conserved separately for the electron and the muon subgroup of the leptons.

LIGHT-YEAR: The distance traveled by light in one year, going at 186,000 miles per second. It equals 5.88×10^{12} miles.

LINES OF FORCE: See FIELD.

LIQUID: A state of matter in which the molecules are in close contact but are free to move about among one another.

MASS: An attribute of all material substance which manifests itself in two universal properties of matter: gravitational attraction and inertia (opposition to change of motion). Newton defined mass as "quantity of matter," meaning, for example, that six identical coins have six times the mass of one. (See also REST MASS.)

MASS FLUCTUATION: The fluctuation of the mass of a particle permitted by the energy-time relations of the Heisenberg uncertainty principle. For a sufficiently brief time interval the permitted fluctuation of mass may be as great as the entire mass of the particle so that it may disappear completely and quickly reappear. Similarly, a particle may appear briefly as a mass fluctuation and quickly disappear again.

MATRIX: An array of symbols arranged in rows and columns representing a system of entities related in a systematical manner. Matrices may be manipulated by the rules of *matrix algebra*, a mathematical method which Heisenberg used in the development of his *matrix mechanics*.

MATTER WAVES: Waves associated with material objects in motion, a concept first proposed by de Broglie. The laws of propagation of these waves determine the mechanical motion of the objects, as described by the theory of *wave mechanics*.

MECHANISM: A theory which asserts that all the phenomena of living organisms may be understood in terms of the physical-chemical laws of inorganic matter. (Compare VITALISM.)

MESON: Any particle having a zero or integral spin value. The masses of the mesons are generally intermediate between the lighter leptons and the heavier baryons. The number of mesons is not conserved in particle interactions. Virtual mesons are the quanta of the strong-force field.

METAPHYSICS: The branch of philosophy which deals with the nature of the physical universe and its cognition by the human mind. (See also EPISTEMOLOGY; ONTOLOGY.)

MILLION-ELECTRON-VOLTS (Mev): A unit of energy, and of equivalent mass, equal to a million (10^6) electron-volts.

MOLECULE: The smallest possible amount of any chemical substance, consisting of a closely bound group of atoms. Molecules of chemical elements consist of atoms of one kind, usually one or two; molecules of chemical compounds are built of two or more kinds of atoms.

MOMENTUM, ANGULAR: The analogous quantity, for rotational motion, to linear momentum for motion along a line. Angular momentum = $I\omega$ where I, the *rotational inertia*, is analogous to m, and ω, the *angular velocity* (rate of turning), is analogous to v. The angular momentum of a particle rotating about its axis is called *spin*. The angular momentum of electrons in atomic orbits is *quantized* in Bohr units. Spins are either integral (quantized in Bohr units) or half-integral (quantized in half Bohr units). (See also SPACE QUANTIZATION.)

MOMENTUM, LINEAR: The product mv of mass and velocity. It is a directed quantity, in the direction of v. At very high velocities, the relativistic increase of mass must be taken into account.

NATURAL NUMBER: Any one of the sequence of integers 1, 2, 3, 4. . . . Natural numbers are of primary importance in the physics of the atomic realm.

NEUTRINO: A stable particle having zero rest mass, zero charge and half-integral spin (a spin of a half Bohr unit).

NODE: A point or a surface in a standing wave pattern at which the wave displacement is zero.

ONTOLOGY: The branch of metaphysics which deals with the nature of the physical universe.

ORGANIC COMPOUND: A chemical compound characteristic of living organisms; more generally, any compound whose molecules are complex structures built of large numbers of atoms, usually including a chain of carbon atoms. (Compare INORGANIC COMPOUND.)

PARITY: A property of particles which is not conserved in weak-force interactions. The nonconservation of parity is tantamount to the existence of processes for which reflection symmetry of space is not valid. (See also REFLECTION SYMMETRY.)

PERIODIC MOTION: Any mode of motion, such as an oscillation or a rotation, which repeats itself in successive equal time intervals.

PERIODIC PROPERTY: Any property of atoms, such as their chemical nature, which occurs repeatedly throughout the orderly sequence of kinds of atoms ranging from the lightest, hydrogen, to the heaviest, uranium. Atomic mass is an example of a nonperiodic property.

PERIODIC TABLE OF ELEMENTS: An orderly arrangement of the chemical elements, in rows and columns, by order of increasing atomic weight (with a few minor exceptions) such that elements having the same chemical properties lie in the same vertical column.

PHASE: The time interval between the instants at which two oscillations or waves attain their maximum displacements. If the maxima coincide, the two waves are said to be *in phase;* if one attains its positive displacement; the other its negative displacement, at the same instant, they are a half-cycle *out of phase.*

PHOTON: A stable particle of radiation. It has zero rest mass, zero charge and a spin of one Bohr unit.

PLANCK'S CONSTANT: The universal constant h of action. It is of primary importance in the physics of atoms and of particles.

POWER, MATHEMATICAL: The number of times a quantity is multiplied by itself. Example: the third power of 2 (written 2^3) is $2 \times 2 \times 2 = 8$; the fourth power of 3 (written 3^4) is $3 \times 3 \times 3 \times 3 = 81$. The second and third powers have the special names *square* and *cube,* respectively.

POWER, MECHANICAL: The rate of performing work. (See also ELECTRICAL QUANTITIES AND UNITS.)

PROTON: The only stable baryon. It has a positive charge of one electron unit (e), a rest mass of 1.672×10^{-24} gram and a half-integral spin.

QUANTIZATION: The restriction of various quantities to certain discrete values. For example, the energy of any system performing a periodic motion of frequency ν can have only the special values $n \times h\nu$; the angular momentum of orbital electrons in atoms can assume only the values of $n \times h/2\pi$, where n is one of the sequence of integers 0, 1, 2, 3. . . . The energy of a photon of frequency ν has only the one value $h\nu$.

QUANTUM ELECTRODYNAMICS: A physical theory dealing with the properties and the interactions of photons, electrons and positrons.

QUARKS: Hypothetical entities which may be combined mathematically to produce the properties of all the mesons and baryons.

RADIANT ENERGY: See ENERGY.

RADIATION: The process whereby the energy inherent in material substance is converted to radiant energy. The reverse process is called *absorption.*

RADIOACTIVITY, ARTIFICIAL: Radioactivity induced in normally nonradioactive kinds of atoms by bombarding their nuclei with protons, neutrons and alpha particles.

RADIOACTIVITY, NATURAL: The spontaneous emission of alpha, beta and gamma rays from the nuclei of atoms, resulting in the transformation of one kind of atom into another, a process called *radioactive decay.* It occurs mostly with atoms near the high atomic number end of the periodic table of elements.

REALISM: See IDEALISM.

REFLECTION SYMMETRY: A property of space which assures that the reflection, as in a mirror, of any possible phenomenon is also a possible one. More generally, this symmetry refers to reflection in all of the three orthogonal directions of space, left-right, up-down and front-back.

REFRACTION: The bending of a light ray away from a straight path when passing from one transparent medium to another, as from air into water or from thinner to denser air.

RESONANCE: The exceptionally strong response of an oscillatory system when it is acted upon by an influence having a frequency equal to the natural frequency of the system. (See also BARYON.)

REST MASS. The mass of an object at rest. When the object is in motion, the mass is increased by the mass equivalent of its kinetic energy. As the velocity approaches that of light, the mass increases greatly toward an infinite value.

SCATTERING: Any phenomenon, such as deflection or creation of new particles, which results from particle collisions.

SOLID: A state of matter in which the molecules are bound together in fixed positions relative to each other but perform small oscillations about these positions because of their heat energy. In crystalline solids the molecules are arranged in an orderly pattern of rows and columns, layer upon layer.

SPACE QUANTIZATION: The restriction of the angular momentum of electrons in atomic orbits and of the spin of particles to certain particular orientations in space. Example: two interacting particles with half-integral spins can have their spins only in the same (parallel) or in opposite (antiparallel) directions.

SPECTRUM: The array of all the wavelengths or colors of light emitted or absorbed by an atom or a molecule.

SPIN: The angular momentum of a particle rotating about its axis. It is a directed quantity oriented in the direction in which the rotation would advance a right-hand screw.

STANDING WAVE: A pattern of oscillations in space in which the regions of maximum displacement and of zero displacement (nodes) remain fixed in position.

STRANGENESS: A property of particles which is significant for the occurrence or absence of some kinds of particle interactions. It is conserved in strong-force but not in weak-force interactions.

STRONG FORCE: The force which binds nucleons together in atomic nuclei and which can bring about particle interactions in the very brief time of about 10^{-23} second.

VECTOR QUANTITY: See DIRECTED QUANTITY.

VELOCITY: Time rate of change of displacement. Velocity is a directed quantity having both magnitude and direction. SPEED refers to the magnitude only.

VIRTUAL PARTICLE: A form in which particles can exist only for the brief time permitted by the mass fluctuations given by the Heisenberg uncertainty principle, a time which is inversely proportional to the mass of the particle. Virtual particles are important as the quanta of force fields.

VITALISM. A theory which asserts that the phenomena of living organisms involve a special "vital principle." (Compare MECHANISM.)

VOLT: See ELECTRICAL QUANTITIES AND UNITS.

VOLUNTARY NERVOUS SYSTEM: The system of nerves which is concerned with all voluntary bodily functions, such as control of musclar action, sense perception and other conscious mental activity. (Compare INVOLUNTARY NERVOUS SYSTEM.)

WAVELENGTH: The distance between successive crests of a wave.

WAVE PACKET: A traveling group of waves confined to a restricted region of space. In the case of matter waves the motion of the wave packet determines the motion of the material object with which it is associated.

WEAK FORCE: A force of unknown nature which produces particle interactions, of particular importance for leptons. It is weaker than the strong force by a factor of about 10^{-13}, and the rate at which it can produce interactions is lower by this same factor.

WORK: See ENERGY.

WORLD LINE: A line in mathematical, four-dimensional space-time which represents the progression of a point both in the three directions of space and in time.

For Additional Reading

The following books, mostly paperbacks, generally do not require more than an elementary knowledge of mathematics and science. They are thoroughly authoritative, interesting and readable. Occasional mathematical discussions may be skipped without appreciable loss of comprehension.

GENERAL PHYSICS AND ATOMIC PHYSICS

ANDRADE, E. N. DA C. *An Approach to Modern Physics.* New York: Doubleday and Company, Inc. (Anchor Books), 1956.

———. *Physics for the Modern World.* New York: Barnes and Noble, Inc., 1963.

These two small books, whose author is a distinguished English scientist and a well-known writer and lecturer, provide a brief survey of classical and modern physics, and a good background for the comprehension of recent developments.

MC CUE, J. J. C., and SHERK, K. W. *The World of Atoms.* New York: The Ronald Press Company, 1963.

A college textbook written for students in the humanities and—says the preface—with the conviction "that science can be taught humanistically without being taught sloppily." The contents, broader than the title implies, include a systematic discussion of basic concepts in physics, chemistry and certain areas of astronomy. The book is well suited for leisurely self-instruction and as a reference.

CHEMISTRY

JAFFE, BERNARD. *Chemistry Creates a New World.* New York: Pyramid Publications, Inc., 1962.

The introduction characterizes the book as "describing for us a dazzling multiplicity of chemical marvels which have revolutionized modern society in the fields of metallurgy, agriculture, medicine and public health, industry, transportation, warfare and in countless other fields." The first three chapters give a brief survey of the history and general principles of chemistry.

MERRILL, P. W. *Space Chemistry.* Ann Arbor, Mich.: The University of Michigan Press, 1963.

About a hundred years ago the philosopher Auguste Comte stated categorically that the chemical composition of the stars is, by its very nature, forever beyond the reach of human knowledge. This book gives an account of how this statement has been refuted through the study of the stars and of the tenuous matter of inter-stellar space by the methods of astrochemistry, including the use of rockets and space vehicles. Transmutation of elements and energy production in the stars are discussed.

NUCLEAR PHYSICS

ROMER, ALFRED. *The Restless Atom.* New York: Doubleday and Company, Inc. (Anchor Books), 1960.

The restlessness discussed here is that of radioactive atomic nuclei; and the book is mainly an account of the discovery and study of radioactivity. The work of Henri Becquerel, the Curies and other early experimenters is discussed in a lively and informative manner.

WILSON, P. R., and LITTAUER, R. *Accelerators: Machines of Nuclear Physics.* New York: Doubleday and Company, Inc. (Anchor Books), 1960.

As implied by the title, this book was written at a time when the primary use of particle accelerators was still in nuclear rather than in elementary particle research. The purpose and use of accelerators is discussed briefly but the book deals mostly with their construction and operation.

ELEMENTARY PARTICLES

FORD, K. W. *The World of Elementary Particles*. Waltham, Mass.: Blaisdell Publishing Co., 1963.

One of the earliest and still one of the best books on elementary particles. The author, an active research scientist, briefly describes experimental techniques and results but deals mainly with ideas and theories.

RELATIVITY

GARDNER, MARTIN. *Relativity for the Million*. New York: Simon and Schuster, Inc. (Pocket Books), 1962.

The author is an editor of *Scientific American* and a highly successful popular science writer. The book includes recent experiments and cosmological theories related to relativity. The remarkable illustrations by Anthony Ravielli help the reader to obtain an intuitive grasp of the subject.

ASTRONOMY AND COSMOLOGY

ALFVEN, HANNES. *Worlds—Antiworlds: Antimatter in Cosmology*. San Francisco: W. H. Freeman & Co., Publishers, 1966.

A leading contemporary astronomer presents his original ideas about the simultaneous creation of matter and antimatter when the universe began.

TOULMIN, S., and GOODFIELD, J. *The Fabric of the Heavens*. New York: Harper & Row, Publishers (Torchbook), 1961.

The authors have drawn upon their wide knowledge of science, history, literature and philosophy to produce an authoritative and intriguing story of the development of ideas about the structure and the dynamics of the astronomical universe, beginning with those of the ancient Babylonians but stopping short of recent happenings in astrophysics, radio astronomy and the use of space vehicles.

NEUROPHYSIOLOGY

ASIMOV, ISAAC. *The Human Brain*. New York: The New American Library (Signet Science Library), 1965.

The author, professor of biochemistry, Boston University School of Medicine, is an eminently successful writer of popular books on science. Ranging wider than the title suggests, the book covers the whole field of internal control in living organisms by hormonal secretions and by the nervous system. This excellent exposition of structure and function, with frequent references to historical developments, requires some honest effort for full appreciation.

HISTORY OF SCIENCE

COHEN, I. B. *The Birth of the New Physics*. New York: Doubleday and Company, Inc. (Anchor Books), 1960.

The "New Physics" discussed here is that of Copernicus, Kepler, Galileo and Newton. This is one of the very best presentations of the birth of classical physics.

RAPPORT, S., and WRIGHT, H. (*eds.*) *Physics*. New York: Simon and Schuster, Inc. (Washington Square Press), 1965.

A collection of eighteen articles by famous physicists and historians of science, discussing outstanding developments in the period from Galileo and Newton up to the present. The editors have provided a general introduction, and separate introductions to the individual articles in which they provide biographical information about each author and relate his topic to the story of the book as a whole.

A Symposium. *A Short History of Science*. New York: Doubleday and Company, Inc. (Anchor Books), 1959.

The sixteen chapters of this book do not form a connected history of science. Rather, they present important advances, in the physical and the biological sciences, which have had a major impact on our concepts about the nature of the universe, starting with the views of Dante Alighieri and proceeding through the Renaissance, the classical period of science, and on into the twentieth century.

TOULMIN, S., and GOODFIELD, J. *The Architecture of Matter*. New York: Harper & Row, Publishers (Torchbook), 1962.

The subtitle, "The Physics, Chemistry and Physiology of Matter, Both Animate and Inanimate, As It Has Evolved Since the Beginnings of Science," aptly characterizes the contents of this book. This tremendous wealth of information is woven into a lively and coherent story with a strong sense of the historic progress of scientific and philosophical concepts from antiquity to the present time.

PHILOSOPHY OF SCIENCE

BEROFSKY, B. (ed.) *Free Will and Determinism.* New York: Harper & Row, Publishers, 1966.

A collection of writings on free will and related subjects, arranged in six chapters, each with an introduction by the editor, who also has supplied a general introduction. With a few exceptions (including St. Augustine), the authors are contemporary philosophers, and some of the writings presuppose competence in this discipline.

DANTO, A., and MORGENBESSER, S. (eds.) *Philosophy of Science.* Cleveland: The World Publishing Company (Meridian), 1960.

This is an anthology of articles by twenty-four European and American philosophers. It is a thick book (470 pages) which provides profitable browsing for anyone interested in the philosophical problems attendant on man's quest for knowledge.

DE BROGLIE, L. *Matter and Light: The New Physics.* (Translated from the French by W. H. Johnston.) New York: Dover Publications, Inc., 1939.

A collection of articles, both scientific and philosophical, on modern physics by the founder of quantum mechanics. The book includes the address given by the author upon receiving the Nobel Prize.

PROBABILITY

WEAVER, W. *Lady Luck.* New York: Doubleday and Company, Inc. (Anchor Books), 1963.

The style of the book is that of an informal conversation; yet it provides a great amount of information about the theory of probability, its history and its many interesting applications in science, metaphysics, technology and everyday affairs. There are a considerable number of mathematical proofs, but their results may be accepted on faith. This book is without doubt the best available source for a general comprehension of the meaning and value of probability theory.

COMPUTERS

FINK, D. G. *Computers and the Human Mind.* New York: Doubleday and Company, Inc. (Anchor Books), 1966.

The author considers technical matters of electronic computer construction and operation but discusses at greater length the broader aspects of computer logic and intelligence, the similarities and differences of computers and human brains and the use of computers in various interesting problems.

GENERAL

DEASON, H. J. (ed.) *A Guide to Science Reading.* New York: The New American Library, Inc. (Signet Science Library), 1966.

Hundreds of additional paperback books in many areas of science are listed, described briefly, and rated in order of difficulty.

Scientific articles in great variety, many of which are pertinent to atomic physics and quantum mechanics, are published in *Scientific American.* Most of these articles are written by scientists doing original research in their several fields; but they are intended to be comprehensible to non-specialists. This magazine is widely available in public, high school and college libraries. Offprints of single articles are published by W. H. Freeman & Company, Publishers, 660 Market Street, San Francisco, California 94104. A catalog is available listing about one thousand offprints by author and subject.

Index

(Page numbers in italics refer to illustrations.)

Illustration Credits

tember, 1964, by permission of W. H. Freeman and Company, Publishers.

Table of quarks (page 196) was abstracted from "Quarkways to Particle Symmetry," by L. M. Brown, *Physics Today*, February, 1966, by permission of *Physics Today*.

Diagram of particle energy levels (page 205) is a simplified version of diagrams appearing in "Quantum Theory and Elementary Particles," by V. F. Weisskopf, *Science*, September 10, 1965, by permission of V. F. Weisskopf and *Science* Magazine. Copyright 1965 by the American Association for the Advancement of Science.

9 780486 428741